Trends in Integrated Insect Pest Management

Edited by R. P. Soundararajan and Chitra Narayanasamy

Published in London, United Kingdom

IntechOpen

Supporting open minds since 2005

Trends in Integrated Insect Pest Management
http://dx.doi.org/10.5772/intechopen.78202
Edited by R. P. Soundararajan and Chitra Narayanasamy

Contributors
Shakti Singh Bhati, Yêyinou Laura Estelle Loko, Dèca Mondoukpè Joelle Toffa, Azize Orobiyi, Hassan-Ali Vahedi, Jabbar Valipour, Abbas Ali Zamani, Zazali Alias, Siti Nasuha Hamzah, Yong Hao Tan, Muhammad Imran, Anicet Dassou, Roland Dossou, Corinne Anagonou, Alexandre Dansi

First published in London, United Kingdom, 2020 by IntechOpen
IntechOpen is the global imprint of INTECHOPEN LIMITED, registered in England and Wales, registration number: 11086078, 7th floor, 10 Lower Thames Street, London, EC3R 6AF, United Kingdom
Printed in Croatia

British Library Cataloguing-in-Publication Data
A catalogue record for this book is available from the British Library

Additional hard and PDF copies can be obtained from orders@intechopen.com

Trends in Integrated Insect Pest Management
Edited by R. P. Soundararajan and Chitra Narayanasamy
p. cm.
Print ISBN 978-1-78984-484-9
Online ISBN 978-1-78984-485-6
eBook (PDF) ISBN 978-1-83880-593-7

Meet the editors

Dr. R. P. Soundararajan, Associate Professor in Horticultural College and Research Institute for Women of Tamil Nadu Agricultural University, Tiruchirappalli, Tamil Nadu, India, has 15 years of experience in research and teaching in Agricultural Entomology. He is specialized in host plant resistance and insect pest management. He completed his Ph.D. on quantitative genetics in host plant resistance of rice crop against brown planthopper. He has published four books, 20 book chapters, and 45 research papers. He has guided 4 postgraduate research students. He is a life member of various academic bodies and active review committee member in several peer-reviewed academic journals.

Dr. Narayanasamy Chitra, Associate Professor & Curator (TNAU Insect Museum) at Department of Agricultural Entomology, Tamil Nadu Agricultural University, Coimbatore, Tamil Nadu, India, has 15 years of experience in research and teaching in Agricultural Entomology. She is specialized in insect taxonomy and biodiversity. She was involved in the establishment of the TNAU Insect Museum. She has published two books and 30 research papers. She has obtained funded research projects related to insect taxonomy from the Department of Science and Technology, Government of India. She is an editorial committee member in two research journals.

Contents

Preface

Insects are the most speciose organism on earth. Insects are adapted to all environments and ecosystems. Insects are one of the major components in the functioning of ecosystems. However, a handful of them pose serious threats to agriculture as pests of crops. In many crops, insects are the major constraint in achieving higher yields, thereby, affecting the productivity of crops. Further, a few species of insects damage the harvested produce in stores. Global estimates reflect a loss of 9.5 billion US$ due to various biological stresses to crop plants of which insect pest damage accounts for 10.8% loss. Apart from causing direct damage to crops, insects also act as vectors of plant diseases, which further decreases crop yields. Apart from being pests, there is a wealth of insects that act as predators, parasitoids, and scavengers. Hence, the knowledge on the role of insects in agro-ecosystems is essential in sustainable crop production systems.

Various management tactics have been deployed to decrease the impact of insects in agriculture since time immemorial. Every time a new technique is deployed, insects have their own techniques to oust the management technique introduced. The most common control measure is the use of chemical insecticides. The impact of the insect pests in field and storage needs to be managed through innovation and sustainable techniques. Integrated pest management (IPM) is popular in combining all available techniques in one capsule to avoid yield losses due to insect damage. IPM techniques are preferred as it provide ways to maintain the environment in balance and provide livelihood for all organisms without reduction in yields. The most striking aspect of IPM is avoidance or minimum use of synthetic chemical compounds. Insecticide application is the last option in any cropping system under this approach. The risk of the 3 R's (development of resistance to insecticides, insect resurgence, and residues) is due to indiscriminate use of pesticides. In addition, there can also be an outbreak of secondary insect pests and destruction of natural enemies.

In the wake of climate change and invasive insects, it is the need of the hour to bring forth insect management techniques based on the basic understanding of how insects would respond to the introduced management techniques. Management through biological control agents, botanical pesticides, natural control, alternate cropping systems, and resistant varieties are a few of the important strategies followed in IPM.

This book deals with different aspects of IPM. Lepidopteran insects are major pests of field and horticultural crop plants. Parasitoids play a major role in controlling lepidopteran insects. An overview of the Hymenopteran parasitoids associated with lepidopteran pests is provided. Entomopathogens to manage *Spodoptera litura* (Noctuidae: Lepidoptera), a polyphagous pest attacking several crops, is detailed. IPM for stored product pests attacking processed yam and maize storage is also covered in the book and different storage structures have been suggested. The effect of the newer insecticide group, neonicotinoids, and its impact on pollinators is explained. The mechanisms of insecticide resistance due to oxidative stress on insects are also discussed. We hope the contents of the book will be useful to the scientific community to widen their understanding of IPM.

We thank all the authors for their valuable contribution in this book. We are indebted to the author service manager, Mr. Josip Knapic, for his sincere, timely correspondence, and his patient approach. Our special appreciation and thanks to the editorial and publication team of IntechOpen for their incessant encouragement, suggestions, and promptness in each stage of compilation of the book.

Dr. R.P. Soundararajan
Associate Professor (Agricultural Entomology),
Horticultural College and Research Institute (Women),
Tamil Nadu Agricultural University,
Tiruchirappalli, Tamil Nadu, India

Dr. Narayanasamy Chitra
Associate Professor (Agricultural Entomology)
and Curator (TNAU Insect Museum),
Department of Agricultural Entomology,
Tamil Nadu Agricultural University,
Coimbatore, Tamil Nadu, India

Section 1

Biological Control

Chapter 1

Lepidopter Parasitoidea

Hassan-Ali Vahedi, Jabbar Valipour and Abbas Ali Zamani

Abstract

Parasitoids have a long history of pest management, specially for control of economical important lepidopteran pests, such as Noctuidae, Tortricidae, and Pyralidae. The two major parasitoids super families Trichogrammatidae and Braconidae in relation to biocontrol of lepidopteron pests are important. In this chapter: (i) the mass production of the moth egg parasitoid, *Trichogramma brassicae* Bezdenko, 1968 (Hymenoptera, Trichogrammatidae), which has been known to occur in Iran and attack many important hosts Lepidoptera, such as the chickpea pod borer, *Helicoverpa armigera* (Hübner, 1808), Carob moth, *Ectomyelois ceratoniae* (Zeller, 1839), and the codling moth, *Cydia pomonella* (Linnaeus, 1758) in the region and (ii) biology and parasitism behavior of *Bracon* (*Glabrobracon*) *variator* Nees, 1811, as larval ectoparasitoid of *Cydia johanssoni* Aarvik and Karsholt, 1993 (Lepidoptera: Tortricidae), a seed feeder of *Anagyris foetida* Linnaeus, 1758 (Fabaceae) at forest habitats in Western Iran, are discussed. Host development was arrested immediately upon parasitism. The dissected capsules show each adult female lay three eggs close to host larvae inside seed case. Video films and photographs of the behavior as research documents were recorded. Both aestivation and hibernation of the parasitoid occur in the parasitization rate on *C. johanssoni* averaged 18.77% ± 3.80, during second generation of the parasitoid wasp, *B.* (*G.*) *variator*.

Keywords: biological agents, Chalcidoidea, ectoparasitoid, Ichneumonoidea, Iran

1. Introduction

Many Lepidoptera larvae are economically important to agriculture and forestry ecosystems. For example, Noctuidae, Pyralidae and Tortricidae. Many Hymenoptera parasitoids, such as Trichogrammatidae and Braconidae are an important potential bio-control agent for control of moth pests. Recently, attention has been focused on this strategy due to increased consumer concern with pesticide residues in food products and a wide-ranging negative impact of chemical insecticides to the environment. Thus, using parasitoid Hymenoptera can be a safe and viable method of crop protection.

2. Hymenoptera parasitoid

Hymenoptera is the third largest and perhaps the most beneficial to humans of all insect orders. It has around 320,000 species, mostly more than 75% (240,000) are parasitoids (in 12 super families). Parasitoid wasps are highly diverse and specialized to attack a particular host life stage (egg, larvae, pupae and adults) of most arthropods, mainly insects. Major species richness of parasitoid wasps is in Ichneumonoidea 100,000 (which include Braconidae 40,000 and Ichneumonidae 60,000 known species) and Chalcidoidea 22,000 known species; Most species of Chalcidoidea are tiny, <3 mm in length; as a result, they can be difficult to collect and study. Detailed study an estimated more than 500,000 species in existence belong to the Chalcidoidea; within this superfamily, Trichogrammatidae are the smallest, ~0.2 mm in length, which includes 83 genera and 839 known species. Some *Trichogramma* species and strains have a wide host range of insect eggs, while others have strong preference for the eggs of a particular moth species and many successful biological control programs have involved the introduction of highly specific parasitoids [1–4]. The two major parasitoids groups, Trichogrammatidae and Braconidae in relation to biocontrol of lepidopteron pests are important. This chapter is concerned mass production of the egg parasitoid, *T. brassicae* Bezdenko, 1968 (Trichogrammatidae); addition refer to biology and parasitism behavior of *B.* (*G.*) *variator* Nees, 1812 (Braconidae), in relation to lepidopteran pests.

3. Trichogrammatidae

Most prominent species of *Trichogramma* are mostly amenable for insectarium mass production (**Figures 1, 2** and **6**) on factitious hosts like the grain moth, *Sitotroga cerealella* (Olivier, 1789), which it selves is mass produced on factory scales and is being used for biological control of noxious Lepidopterous pests of crops worldwide (**Figures 1, 2** and **6**). *Trichogramma* adults are typically free-living and the females are responsible for finding host insects for their progeny.

3.1 *Trichogramma* biology

The development of all *Trichogramma* spp. is very similar. The eggs hatch in about 24 h and the parasite larvae develop through three instars. These are followed by a prepupa, when the adult characters form, and a pupa. At the beginning of the third larval instars, the host egg turns black due to the deposition of black granules at the inner surface of the chorion, an invaluable diagnostic character for parasitized eggs. After about 5 days, the adult wasps emerge from the pupae and escape the bollworm egg by chewing a circular hole in the egg shell (**Figure 2**). *Trichogramma* overwinter as immature forms in host eggs. This short life cycle allows multi generations per year, and rapid population increase [5]. Hence, early season releases

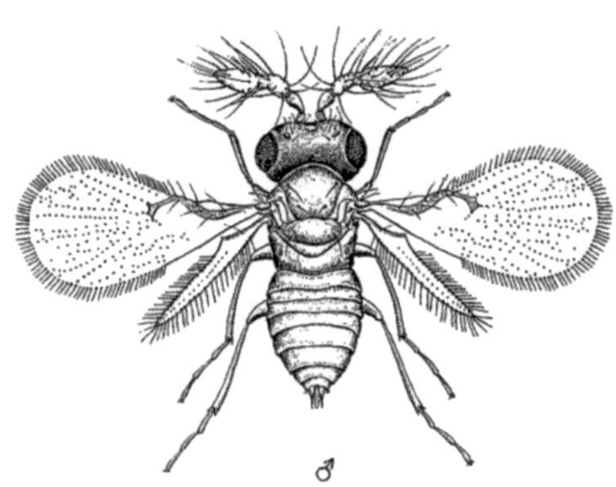

Figure 1.
General view drawing of the moth egg parasitoid, Trichogramma australicum Girault, 1912 (Hymenoptera: Trichogrammatidae), adult male ×120 [6].

DOI: http://dx.doi.org/10.5772/intechopen.86381

produce large populations positioned to fight pest invasions. Unparasitized eggs remain light until black at hatching stage (**Figure 3**). For example, the development time for *T. australicum* Girault, 1912 (Hymenoptera: Trichogrammatidae) on the rice moth, *Corcyra cephalonica* (Stainton, 1866) (Lepidoptera: Galleriidae) at 28 ± 2 from egg to adult are completed in 14 days [5, 6]. The early stages of developed eggs (**Figure 3B**) are more suitable for parasite development. Older bollworm eggs, especially those in which the head capsule of the larva is visible, are not usually parasitized and if they are, parasite survival is much lower (**Figure 3C**) [7].

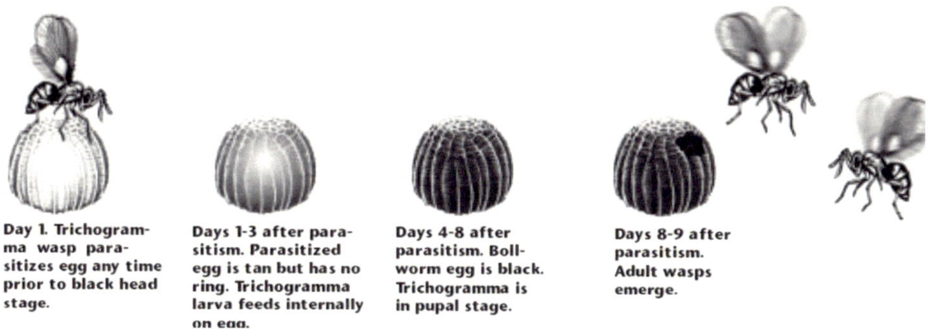

Figure 2.
Life cycle of the moth egg parasitoid, Trichogramma developing on the bollworm egg (Helicoverpa armigera) (Hübner, 1808) [5] (moth egg diameter 0.4–0.6 mm).

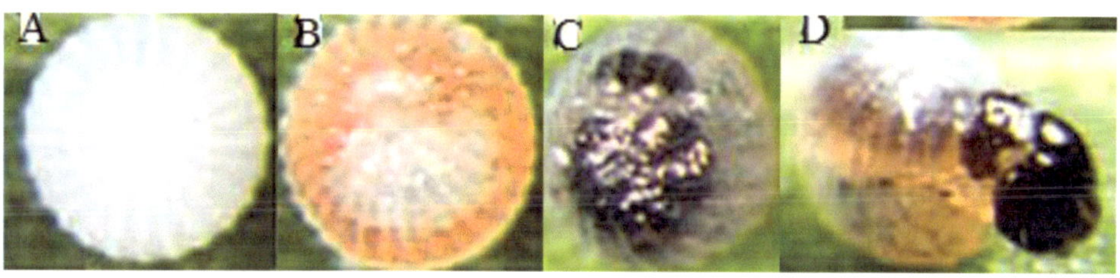

Figure 3.
Light micrographs of development of the pod borer eggs (unparasitized), Helicoverpa armigera (Hübner, 1808) showing the embryonic developmental sequences: (A) newly laid egg (yellowish-white); (B) 2 days old egg (tan egg); (C) egg before hatching or black head stage (larva visible) and (D) hatching stage (larva) (moth egg diameter 0.4–0.6 mm). Original.

Figure 4.
Images general view of the chickpea pod borer, Helicoverpa armigera (Hübner, 1808) (Lepidoptera: Noctuidae): (Right) 5th larval stage off the chickpea and (Left) The adult stage (Lepidoptera: Noctuidae); (adult wingspan 3.5–4 cm; 1.5–1.9 cm long and the mature larva or last instars was about 2.8 ± 0.05 mm long and 0.27 ± 0.08 cm wide). Original.

Figure 5.
Images general view of the pomegranate calyx worm, Ectomyelois ceratoniae (Zeller, 1839) (Lepidoptera: Pyralidae) and infested pomegranate fruit: Right to left. The adult stage of E. ceratoniae; 5th larval stage of E. ceratoniae off the pomegranate; infested split pomegranate fruit and applied cardboard capsule for infested orchards; (adult wingspan 1.6–2.5 cm, 1.5–1.8 cm long and the mature larva or last instars was about 1.8 ± 0.07 cm long and 0.75 ± 0.05 cm wide). Original.

Recently, mass rearing techniques have been developed for several parasitoids of insects pests, including, *Trichogramma* (Chalcidoidea: Trichogrammatidae). Currently, several laboratories are actively engaged in producing *Trichogramma* on local scale. These biological agents, parasitoids, are mass produced and released for controlling variety of insect pests including, the chickpea pod borer, *H. armigera* (Hübner, 1808) and the Pomegranate calyx worm, *E. ceratoniae* (Zeller, 1839), which maintain populations below economic levels [8, 9]. The chickpea pod borer, *H. armigera* (Hübner, 1808), (**Figure 4**) produces 52.5% losses [10] and more than 60% losses in the chickpea grower area every year [11]. It attacks on other Legume variety of economically important crops.

The Carob moth, *E. ceratoniae*, which known "Pomegranate calyx worm" (**Figure 5**). It is a pest that attack on variety of fruit crops such as Pomegranate, *Punica granatum* Linnaeus, 1880; Apples, *Malus domestica* Borkh, 1803; date palm, *Phoenix dactylifera* Linnaeus, 1753; Cultivated pistachio, *Pistacia vera* Linnaeus, 1753 [12, 13]. Split pomegranate fruit are more prone to infestation by larva as penetration into the fruit occurs more readily. The adult female deposits its eggs in already split fruits or on the skin or calyx of pomegranate fruit. It is widely distributed and occurs in Iran and close countries [12].

4. *Trichogramma brassicae* Bezdenko, 1968 (Hymenoptera: Trichogrammatidae)

The moth egg parasitoid, *T. brassicae* Bezdenko, 1968 is reared in private or government owned insectaries and released annually in the agricultural crops and orchards in Iran [14]. For example, the development time for *T. brassicae* Bezdenko, 1968, on the grain moth, *Sitotroga cerealella* Olivier, 1789 (Lepidoptera: Galleriidae) at 28 ± 2°C from egg to adult is completed in 14 days [14].

T. brassicae Bezdenko, 1968, pupae can be programmed to enter an overwintering condition of arrested development called diapauses. Once in diapauses, wasp pupae can be stored for up to 9 months so that the large demand for *Trichogramma* during the summer can be met [14, 15].

Cardboard capsules containing host eggs with developing *Trichogramma* are applied to release of *Trichogramma* in the chickpea farms or pomegranate orchards (**Figure 5**).

Released *Trichogramma* are at different developmental stages so that adults emerge from the capsules over several days. This increases the time interval between applications.

Two releases each at a rate of 460,000 pupae per hectare are made beginning at the first moth flight as determined by light traps. Chickpea pod borer, eggs hatch after about 5–6 days and the egg-laying period continues for 3–6 weeks. In-field reproduction of released parasites is believed to be important in providing residual control of eggs deposited after the second release. Field evaluations in Germany have shown releases result in a 70–93% reduction in corn borer larvae relative to untreated fields [15].

In western parts of Iran, releases of *Trichogramma* are a parts of integrated pest management in controlling the chickpea pod borer, *H. armigera* (Hübner, 1808) and the Carob moth, *E. ceratoniae* (Zeller, 1839).

Parasitoid attributes include: The ability to parasitize and develop in the target host egg, the species' preference for the target host egg, total egg mortality caused by parasitism, adult feeding, fecundity, development rate, sex ratio, and longevity releases [16, 17]. These characters are important in mass-rearing programs and then field releases.

5. Parasitism behavior of *Trichogramma*

Trichogramma spp. are most famous biocontrol agents and widely distributed in the world. *Trichogramma* drills a hole through the egg-shell and inserts two to three eggs into eggs of 200 pest moth species, including *Helicoverpa* spp., *Chilo* spp. the pink bollworm (*Pectinophora gossypiella* Saunders, 1844 and etc.) and preventing neonate larvae from hatching out and devouring crops. These parasitoids wasps are so small, <1 mm long; moth egg size, and hence how many of their own eggs to lay, is calculated by timing walks across moth egg surfaces. *Trichogramma* larvae eat out the insides of pest eggs, pupate, and cut an exit hole in moth eggshells for winged adults to squeeze through. Males emerge first, wait for females, and immediately mate.

6. Materials and methods: Mass production technique

Native parasitoid has proved to be one of the most potent egg parasitoid for so many important caterpillar pests. It is reared on factitious hosts such as, the flour moth, *Ephestia kuehniella* Zeller, 1879 (Lep.: Pyralidae), the grain moth, *S. cerealella* (Olivier, 1789), (Lep.: Gelechiidae) and other stored grain pests. The eggs of above hosts are utilized in mass rearing of this parasitoid [18]. *Sitotroga* eggs, test tubes, egg cards, refrigerator, UV chamber (fluorescent tube light, 15 W/Luminal flow), honey solution 20%, gam arabic, camel hair brush, Glass and adults of the grain moth, *S. cerealella* (Olivier, 1789), (Lep.: Gelechiidae). Following rearing procedure are involved in mass production of *T. brassicae* Bezdenko, 1968 (**Figure 6**).

Providing the Grain moth eggs: (i) For prevention from hatching of the eggs during and after parasitization of eggs, they should be exposed to UV rays (15 watt UV tubes) for 45 min at a distance of 5 cm.

Parasitoid species name, date of release of parasitoid for parasitization, expected date of emergence and Institute name and name of technical person should be given on the each egg sheet.

Processed egg sheet is than placed the plastic box. Introduce 6 days old duly parasitized egg cards to adult females of *Trichogramma* for 24 h. The parasitized and unparasitized eggs in each of box/containers should be in the ratio of 1:6 (parasitoid:host) to have optimum parasitization. Close the container properly for preventing escape of parasitoids. The parasitoids emerged in the container will parasitize the unparasitized eggs of the grain moth, *S. cerealella* (Olivier, 1789).

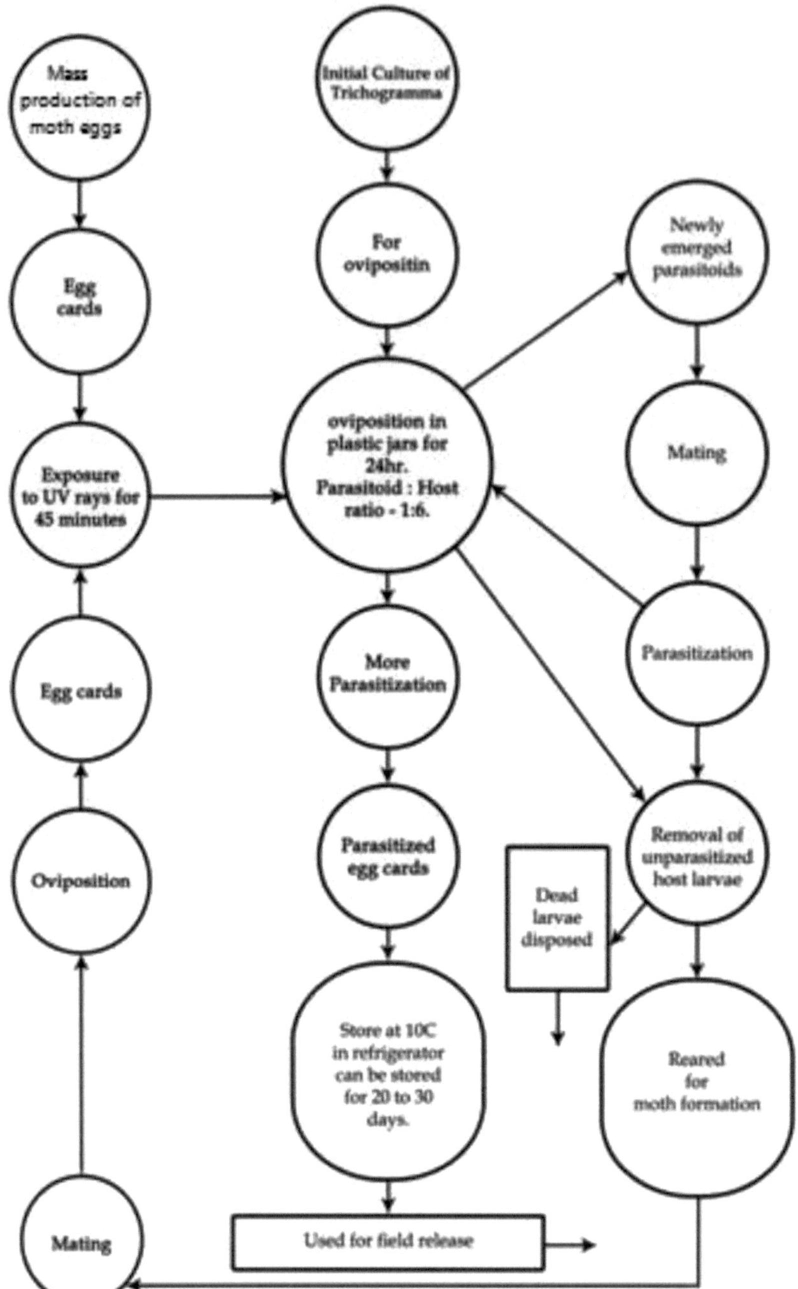

Figure 6.
Mass rearing procedure schematic: conventional production of egg parasitoid, Trichogramma brassicae, and its laboratory host, the eggs of grain moth, Sitotroga cerealella (Olivier, 1789).

Approaching the time of emergence of parasitoid, the egg color becomes blackish. Mating of sexes is essential for obtaining sex ratio, male:female, 1:1. Thus, mating is occurred in container.

Unparasitized eggs get hatched, into larvae (of host), such larvae should be removed from the container with the help of camel hair brush. The host larvae may destroy the eggs hence they should be removed from the container. The parasitoid complete its life cycle with 7–8 days.

After 4 days of parasitization, parasitized eggs of the grain moth, *S. cerealella* (Olivier, 1789), start changing their color from creamy white to blackish due to accumulation of urate granules. Such egg sheets are removed and stored in refrigerator at 10°C temperature for about 20–30 days for our convenience, duskiness point of view. The egg sheets thus preserved can be marked to farmers.

7. Method of calculation of parasitism rate and observation of oviposition behavior of *Bracon* (*Glabrobracon*) *variator* Nees, 1812, (Hymenoptera: Braconidae)

In order to determine parasitism rate of the ectoparasitoid, *B.* (*G.*) *variator* Nees, during first third of June, at least 100 infested seed pods (the stinking bean trefoil, *Anagyris foetida* Linnaeus, 1758 (Fabaceae)) by larva of *Cydia johanssoni* (Lepidoptera: Tortricidae) were collected daily and the number of parasitized larvae was counted and parasitism rate was calculated. The parasitoid was visible on the outside of the caterpillars.

In order to observe parasitism behavior, several live female wasps were collected along with pods containing unparasitized larvae and transported to laboratory. One adult female wasp was released into a petri dish with an infested pod and oviposition observed.

A digital camera, model Canon, PowerShot S3 IS and a digital Sony EXwave HAD camera connected to a stereomicroscope, model Blue Light, were used to take video films and photographs of the wasp behavior.

Immediately after oviposition, the seed capsules were dissected, under a stereomicroscope to observe the number of parasitoid eggs laid.

8. Braconidae

The Braconidae constitute one of the most species-rich hymenopterous families, with 48 subfamilies, more than 1050 genera and about 17,600 species described worldwide. The family has two major lineages: (a) the cyclostome braconids, most of which are idiobiont ectoparasitoids of concealed Lepidoptera and Coleoptera larvae although many are koinobiont endoparasitoids of Diptera and Hemiptera and (b) the noncyclostome braconids which are all endoparasitoids and are generally koinobionts, typically attacking the early instars of their hosts [19, 20].

Taxonomic significance of the genus *Bracon* Fabricius, 1804: The width of the hypoclypeal depression may be important at subgenera level; presence or absence of occipital carina and shape of first metasomal tergite (t_1) (flattened or not flattened lateral parts) [21]. *Bracon* genus is measured as a paraphyletic or also a polyphyletic group, mainly of small and middle-sized species [22] divided over several subgenera [23, 24]. For example, *Bracon* fauna of Iran include five subgenera: *Bracon* (*Bracon*) kozak Telenga, 1936; *B.* (*Glabrobracon*) Fahringer 1927; *B.* (*Habrobracon*) Ashmead, 1895; *B.* (*Asiabracon*) Tobias 1957 and *B.* (*Orthobracon*) Fahringer, 1927 [25]. Braconids also vary greatly in their biology [26–28].

Different species of parasitoids attack different life stages of the pest. For example, *B.* (*G.*) *variator* Nees, 1812, are larval ectoparasitoid and prefers 3rd–5th instars larvae and each adult female laid three eggs close to host larvae (**Figure 7**), inside seed case [32]. Most parasitoids behaviors are so identical and specialized that they can attack only a particular host life stage i.e., they attack eggs (eggs parasitoids), larvae (larval parasitoids), Pupae (pupal parasitoids), or adult (adult's parasitoids) [1]. There are considerable variations in insect parasitoids parasitism. These may be

Figure 7.
Image general view of the larval ectoparasitoid wasp, adult female, Bracon (Glabrobracon) variator Nees, 1812 (Hymenoptera: Braconidae). Original.

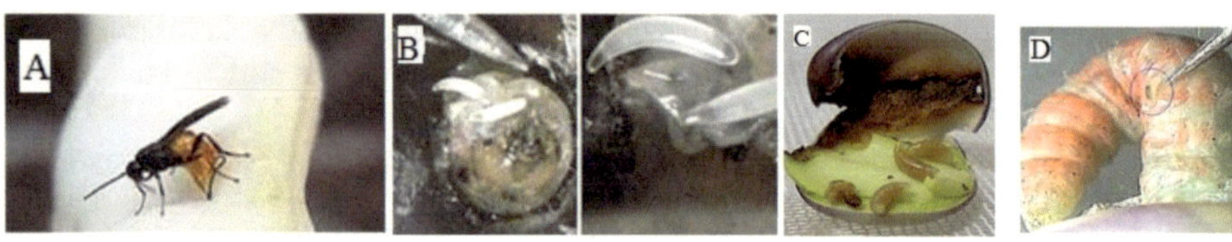

Figure 8.
Images general view of the larval ectoparasitoid wasp, Bracon (Glabrobracon) variator Nees, 1812, (Hymenoptera: Braconidae) and its host larva, Cydia johanssoni Aarvik & Karsholt, 1993 (Lepidoptera: Tortricidae): (A) female, oviposition on infested pod of stinking bean trefoil by pyralid seed feeder larva; (B) parasitoid's eggs ((B. (G.) variator)) laid close to the head of host larva (C. johanssoni); (C) developed parasitoid larvae inside the seed and (D) parasitoid larvae ((B. (G.) variator)) feeding upon Pyralid seed feeder larva, Cydia spp. [31].

idiobiont, whose hosts stop development, when they are parasitized. Idiobionts are either ectoparasitoids that kill their hosts or endoparasitoids that attack immobile host stages such as eggs or pupae. Koinobionts, allow the hosts to continue their development until the parasitoid's offspring matures. Most koinobionts are endoparasitoids of larval stages of insects, although a few are ectoparasitic [27]. Some female parasitoids also use the ovipositor to puncture a host and then feed on the body fluids before selecting other hosts for oviposition, thus causing two different types of mortality in the caterpillar pest population. In some cases, for example, *B.* (*G.*) *variator* Nees, 1812, the egg is laid externally on the body of the host and the larvae also feed externally. This parasitoid wasp is a highly polyphagous gregarious ectoparasitoid that attacks the larvae of a wide range of insects, such as Lepidoptera (**Figure 8**). For example, see [28], *B. (G.) variator* Nees, 1812, paralyzes the larvae of *Hadena bicruris* (Hufnagel, 1766) (Lepidoptera: Noctuidae) before depositing on average 3 eggs on 3rd-5th larval instars [29, 30]. Like *C. johanssoni* Aarvik & Karsholt, 1993, *H. bicruris* (Hufnagel, 1766) is a seed specialist, feeding on *Silene latifolia* Poir. 1789 (Caryophyllaceae) seeds. Lepidoptera larvae, attacks the young seeds before they are shed by the parent plant [32–34]. The larvae of the beech moth, *Cydia fagiglandana* (Zeller, 1841), feed inside the nuts of European beech, *Fagus sylvatica* Linnaeus, 1753 (Fabaceae) causing high seed mortality in South Sweden beech forests [35, 36]. Other economically important species are the pea moth, *C. nigricana* (Fabricius, 1794), attacking legume crops; the spruce seed moth, *C. strobilella* Linnaeus, 1758, attacking spruce seed, *Picea* spp.; *C. latiferreana* Walsingham, 1879, attacking fruits oak and the hickory shuckworm moth, *C. caryana* Fitch, 1756, an important pest of pecan [37].

First generation adult parasitoids of *B.* (*G.*) *variator* Nees, 1812, appeared in the first ten days of May, with the second generation appearing about a fortnight later, but adult wasps were most abundant, and its percentage parasitism were highest among larvae, in early June (**Table 2**). Adult activity was greatest during the hottest part of the day (12.00–14.00 hours). The parasitoid population was greatest on pods infested with *C. johanssoni* Aarvik and Karsholt, 1993, larvae, early June. Oviposition began after a lengthy search period. The female wasp first inspected the drilling into the pod (**Figure 8A**), using her antennae by tapping the pod. She then moved so that her mesothoracic legs straddled the drilling position, lifted her abdomen and inserted her ovipositor almost vertically into the pod (recorded Video film). Almost always each caterpillar had three eggs laid on it (**Figure 8B**); rarely, it was noted that more than 3 eggs were laid up to a maximum of six.

The oviposition is not an easy task. The tip of the ovipositor almost always gets stuck to tiny irregularities of the pod surface. When she loses her balance, re-starts oviposition from the beginning. Since the exact point of drilling is crucial and must be recalculated for accuracy. The eggs were white and bacilli-form, with a diameter approximately equal to the tip of a lab needle (**Figure 8B**). Eggs were oviposited directly onto the host larvae. The incubation period is 18 hours at 33°C and relative humidity of 14%. Upon hatching, the parasitoid larva penetrated the caterpillar's cuticle and fed on the body hemolymph for 4–5 days. Upon seed maturation, the parasitoid maturated larva secreted a webbed cocoon within the seed capsule and then pupated. Adults of second generation leave the seed pod in last ten days of May via a hole made by adult. Adults lived as free-living adults for up to 3–4 days until they mated and oviposited. When the larvae of *C. johanssoni* were paralyzed, their movement became reduced. As a result of feeding by the parasitoid larvae, the moth larva was weakened and eventually is reduced to the head capsule and body cuticle.

The free living adult wasps fed on resin produced when opens the first larval stage of *C. johanssoni* Aarvik and Karsholt, 1993, in the pod. The parasitoid aestivation and hibernation was as a pupa in a cocoon inside infested seeds. There are two overlapping generations during a year in natural conditions because of the diapause the parasitoid larvae go into at the end of the second generation. Additionally, parasitism activity of *B.* (*G.*) *variator* Nees, 1812, was also observed on an unknown seed feeder (Lepidoptera: Pyralidae). This was the only other host record for *B.* (*G.*) *variator* Nees, 1812. The unknown pyralid larva was green in color with large body, larger than *C. johanssoni* Aarvik and Karsholt, 1993, larva and had a very low population in the study area (**Figure 8D**).

The parasitization rate on *C. johanssoni* averaged 18.77% ± 3.80, during second generation of *B.* (*G.*) *variator* Nees, 1812 (**Tables 1** and **2**) [31].

General distribution of this wasp includes: China, Central Asia, Mongolia, Siberia, Russia, Crimea, Iran, Turkey and European country [26].

Stages	Egg	1st–3rd larval instars	Prepupa	Pupa	Adult longevity	Total	Release period	Sex ratio M/F
Day/s	1	4	1	2	7	14	1	1:1

M = male and F = female.

Table 1.
The life stages longevity, release period and sex ratio of Trichogramma brassicae Bezdenko, 1968, reared on the grain moth eggs, Sitotroga cerealella Olivier, 1789 (Lepidoptera: Galleriidae) at 28 ± 2°C.

Date	Total number of *C. johanssoni* larvae	Number of parasitized larvae of *C. johanssoni* by *B. variator*	Parasitism rate (%)
1 June 2014	120	9	7.5
2 June 2014	105	7.5	7.14
3 June 2014	120	9	7.5
4 June 2014	120	7	5.83
5 June 2014	135	22	16.29
6 June 2014	110	20	18.18
7 June 2014	130	29	22.3
8 June 2014	110	38	34.54
9 June 2014	100	33	33
10 June 2014	110	39	35.45
Mean	116	21.35	18.77 ± 3.80

Table 2.
Total number, number of parasitized larvae and the percentage parasitism of Cydia larvae (Cydia johanssoni Aarvik and Karsholt, 1993 (Lepidoptera: Tortricidae) by the second generation of Bracon (Glabrobracon) variator *Nees, 1812 (Hymenoptera: Braconidae), Iran, 2014.*

9. Discussion

Trichogramma brassicae Bezdenko, 1968 (Hymenoptera, Trichogrammatidae) has been the object of great interest regarding its mass rearing and is used as a bio-control agent against many moth pests, in studied area. This study also confirmed that *B.* (*G.*) *variator* is a gregarious idioboint ectoparasitoid with 2 generation in a year, attacks third-fifth instars *C. johanssoni* Aarvik and Karsholt, 1993, caterpillars. Almost always 3 parasitoid eggs were oviposited on each host larva within the pod, which had been previously paralyzed. These observations agree with those of Elzinga [29]. However, Elzinga has not mention about number of eggs and location of pupation.

Because of proper parasitism rate of *B.* (*G.*) *variator* Nees, 1812, this parasitoid plays important role in decrease population of *C. johanssoni* Aarvik and Karsholt, 1993, larvae and it is the most important natural enemy of *C. johanssoni* Aarvik and Karsholt, 1993, in Iran, therefore with conservation of this wasp would take an important step toward reduction of damage of pest moth and development of Stinking bean trefoil shrub in the west of Iran.

Parasitoid sex ratio: In the field condition, female population of *B.* (*G.*) *variator* Nees, 1812, was much more than male and male population was rare.

B. (*G.*) *variator* Nees, 1812, has been recorded from most parts of Iran, in East Azerbaijan [38], South Iran [25], North Central Iran [39] and from western part of Iran [31]. *B.* (*G.*) *variator* Nees, 1812, is already known as a larval parasitoid of the lychnis, *Hadena bicruris*, (Hufnagel, 1766), which is the most important of *Silene latifolia* Poir. 1789 (Caryophyllaceae). This ectoparasitoid attacks 3rd-5th instars moth larvae and stops host development immediately by paralyzing the caterpillars. *B.* (*G.*) *variator* Nees, 1812, is a gregarious parasitoid, which means that several parasitoid larvae attack a given caterpillar and produces clutches that are predominantly single-sex, mainly female (each individual parasitoid lays predominantly three sexed eggs). Females inject paralyzing venom into the host before oviposition. On hatching, the parasitoid larvae perforate the cuticle and imbibe nutrients from the paralyzed or dead host. The lychnis, *Hadena bicruris*, (Hufnagel, 1766) hibernates as a pupa in a cocoon [29]. The adult wasp *B.* (*G.*) *variator* Nees, 1812, feeds on resin of infested pods in the studied forest habitat, western Iran. *B.* (*G.*) *variator* Nees, 1812, is a widespread species, known from: China, Middle Asia, Mongolia, Siberia, Russia, Crimea, Iran, Turkey and Europeans countries [40].

Author details

Hassan-Ali Vahedi*, Jabbar Valipour and Abbas Ali Zamani
Department of Plant Protection, Compus of Agriculture and Natural Resources, Razi University, Kermanshah, Iran

*Address all correspondence to: vnassah@yahoo.com

References

[1] Pennachio F. Strand MR evolution of developmental strategies in parasitic Hymenoptera. Annual Review of Entomology. 2006;**51**:233-258

[2] Noyes JS, Valentine EW. Chalcidoidea (Insecta: Hymenoptera)—Introduction, and review of genera in smaller families. Fauna of New Zealand. 1989;**18**:1-91

[3] Walter S. The biology and ecology of egg parasites of the genus *Trichogramma* Westwood (Hym., Chalc.). Part I.: Studies in selected forest biocoenoses in the G.D.R. Zoologische Jahrbücher, Abteilung für Systeamtik, Ökologie und Geographie der Tiere. 1983;**110**(3):271-299 (in German)

[4] Walter S. Biological and ecological studies on egg parasites of the genus *Trichogramma* Westwood (Hym. Chalc.). Part 2. Investigations carried out under laboratory conditions. Zoologische Jahrbücher, Anteilung für Systematik, Ökologie und Geographie der Tiere. 1983;**110**(4):419-441 (in German)

[5] Knutson A. The Trichogramma Manual: A Guide to the Use of Trichogramma for Biological Control with Special Reference to Augmentative Releases for Control of Bollworm and Budworm in Cotton B-6071. College Station, TX: Texas Agriculture Extension Service, Texas A&M University System; 1998

[6] Metcalfe JR, Brenière J. Egg parasites (*Trichogramma* spp.) for control of sugar cane moth borers. In: Williams JR, Metcalfe RW, Mungomery RW, Mathes R, editors. Pests of Sugar-Cane. Amsterdam: Elsevier Publishing Company; 1969. pp. 81-116

[7] Ruberson JR, Kring TJ. Parasitism of developing eggs by *Trichogramma pretiosum* (Hymenoptera: Trichogrammatidae): Host age preference and suitability. Biological Control. 1993;**3**:39-46

[8] Jozian A, Baigi S. Efficiency control of parasitoid, *Bracon hebetor* (Say) (Hymenoptera: Braconidae) on the chickpea pests. In: 5th Iranian Pulses Symposium, Ilam Province; 2012. Available from: https://www.civilica.com/Paper-PULSES05-PULSES05_028.html.COI:PULSES05_028

[9] Salehnia N. Biocontrol of the pomegranate calyx worm, *Ectomyelois ceratoniae* (Lepidoptera, Pyralidae) in pomegranate orchards. In: 4th National Symposium on Pomegranate, Iran; 2011. Ferdos: Khorasan. Available from: https://www.civilica.com/Paper-ANAR01-ANAR01_083.html [in Persian]

[10] Sevcan O. Natural parasitism rate of *Trichogramma evanescens* Westwood (Hymenoptera: Trichogrammatidae) and its release efficacy against the cotton bollworm, *Helicoverpa armigera* Hübner (Lepidoptera: Noctuidae) in the Cukurova region Turkey. Entomological News. 2008;**119**(1). DOI: 10.3157/0013-872X(2008)119[19,NPROTE]2.0.CO;2

[11] Rabieh MM. Biodiversity of noctuid moths (Lepidoptera: Noctuidae) in the agroecosystems of Mashhad County. Biodiversity International Journal. 2018;**2**(2):147-151. DOI: 10.15406/bij.2018.02.00057

[12] Mehrnejad MR. The current status of pistachio pests in Iran. In: Ak BE, editor. XI GREMPA Seminar on Pistachios and Almonds. CIHEAM. 2001;**56**:315-322

[13] Mehrnejad MR. The carob moth, a pest of pistachio nut in Iran. Acta Horticulturae. 1995;**419**:365-372

[14] Vahedi HA. Mass production of insect pests' natural enemies as biological control agents (Trichogramma and Cryptolemus). Unpublished manuscript. Kermanshah, Iran: Department of Plant Protection, Campus of Agriculture and Natural Resources, Razi University; 2012; 30 pp

[15] Bigler F. Quality control in Trichogramma production. In: Wajnberg E, Hassan SA, editors. Biological Control with Egg Parasitoids. Oxon, UK: CAB International; 1994. pp. 93-111

[16] Vasquez LA, Shelton AM, Hoffmann MP, Roush RT. Laboratory evaulation of commercial Trichogrammatid products for potential use against *Plutella xylostella* (L.) (Lepidoptera: Plutellidae). Biological Control. 1997;**9**:143-148

[17] Hassan SA. Strategies to select *Trichogramma* species for use in biological control. In: Wajnberg E, Hassan SA, editors. Biological Control with Egg Parasitoids. Oxon, UK: CAB International; 1994. pp. 55-71

[18] Nagaraja H. Mass production of Trichogrammatid parasitoids. In: Sithanantham S, Ballal C, Jalali S, Bakthavatsalam N, editors. Biological Control of Insect Pests Using Egg Parasitoids. New Delhi: Springer; 2013

[19] Lachaud JP, Gabriela PL. Diversity of species and behavior of Hymenopteran parasitoids of ants: A review. Psyche. 2012;**2012**:1-24. Article ID 134746

[20] Shaw MR, Huddleston T. Classification and biology of braconid wasps (Hymenoptera: Braconidae). In: Dowling WR, Askew RR, editors. Handbooks for the Identification of British Insects, Part 11. Vol. 7. London, UK: Royal Entomological Society of London; 1991. pp. 1-126

[21] Ward DF. Braconidae of New Zealand. 2014. Available from: http://braconidae.landcareresearch.co.nz [Accessed: 15 March 2019]

[22] Quicke DLJ. The Braconid and Ichneumonid Parasitoid Wasps: Biology, Systematics, Evolution and Ecology. Oxford, UK: Wiley-Blackwell; 2015. 704 pp

[23] Belokobylskij SA, Tobias VI, Kotenko AG, Proshchalykin M, Yu DSK. 48. Fam. Braconidae – Braconidy. In: Lelej AS, editor. Annotated Catalogue of Insects of the Russian Far East. Vol. I Hymenoptera. Vladivostok: Dal'nauka; 2012. pp. 300-389

[24] Broad GR, Shaw MR, Godfray HC. Checklist of British and Irish Braconidae (Hymenoptera). 2012. Available from: http://www.nhm.ac.uk/resources-rx/files/braconidae-checklist-for-web-34139.pdf

[25] Ameri A, Talebi AA, Beyarslan A, Kamali K, Rakhshani E. Study of the genus *Bracon* Fabricius, 1804 (Hymenoptera: Braconidae) of Southern Iran with description of a new species. Zootaxa. 2013;**3754**(4):353-380

[26] Yu DS, van Achterberg C, Horstmann K. World Ichneumonoidea 2004. Taxonomy, biology, morphology and distribution, Taxapad 2005 (Scientific names for information management), interactive catalogue on DVD/CDROM; Canada: Vancouver; 2012

[27] Sharkey MJ. Family Braconidae, in Hymenoptera of the world: An identification guide to families. In: Goulet H, Huber JT, editors. Ottawa, Canada: Research Branch, Agriculture Canada Publications; 1993. pp. 362-395

[28] Quicke DLJ, van Achterberg C. Phylogeny of the subfamilies of the family Braconidae (Hymenoptera:

Ichneumonoidea). Zoologische Verhandelingen Leiden. 1990;**258**:1-95

[29] Elzinga JA. The effects of habitat fragmentation on a tri-trophic system: *Silene latifolia*, *Hadena bicruris* and its parasitoids [thesis]. Netherlands: Utrecht University; 2005. 184 pp

[30] Elzinga JA, Nouhuys S, Leeuwen DJ, Biere A. Distribution and colonization ability of three parasitoids and their herbivorous host in a fragmented landscape. Basic and Applid Ecology. 2005;**14**:1-14

[31] Valipour J, Vahedi HA, Zamani AA. An outline on biology and behavior of *Bracon variator* Nees, 1812 (Hym.: Braconidae), an ectoparasitoid of *Cydia johanssoni* Aarvik and Karsholt, 1993 (Lepidoptera: Tortricidae) from Iran. Biharean. Biologist. 2017;**11**(1):15-19

[32] Valipour J, Vahedi HA, Zamani AA. Preliminary morphological study of developmental stages of *Cydia johanssoni* Aarvik and Karsholt (Lep: Tortricidae) on the stinking bean trefoil, *Anagyris foetida* L. (Leguminosae) in Iran. Journal of Applied Environmental and Biological Sciences. 2015;**5**(7):299-304

[33] Ramirez N, Traveset A. Predispersal seed-predation by insects in the Venezuelan central plain: Overall patterns and traits that influence its biology and taxonomic groups. Perspectives in Plant Ecology, Evolution and Systematics. 2010;**12**(3):193-209

[34] Zhang J, Drummond FA, Liebman M, Hartke A. Insect predation of seeds and plant population dynamics. Maine Agricultural and Forest Experiment Station Technical Bulletin. 1997;**163**

[35] Nielsen BO. Beech seeds as an ecosystem component. Oikos. 1977;**29**:268-274

[36] Nilsson SG, Wastljung U. Seed predation and cross pollination in mast seeding beech (*Fagus sylvatica*) patches. Ecology. 1987;**68**:260-265

[37] Oboyski PT. The Systematics, Evolution, and Ecology of Hawaiian Cydia (Lepidoptera: Tortricidae) [thesis]. Berkeley: University of California. 2011; 123 pp

[38] Rastegar J, Sakenin H, Khodaparast SH, Havaskary M. On a collection of Braconidae (Hymenoptera) from East Azarbaijan and vicinity, Iran. Calodema. 2012;**226**:1-4

[39] Zargar M, Talebi AA, Hajghanbar H, Papp J. A study on the genus *Bracon fabricius* (Hymenoptera: Braconidae) in North Central Iran with four new records for Iranian fauna. Entomofauna: Zeitschrift fur Entomologie. 2015;**36**(32):425-440

[40] Guler Y, Cagatay N. Systematical studies on the genus *Bracon (Glabrobracon)* (Hymenoptera, Braconidae: Braconinae) in Ankara province. Turkish Journal of Zoology. 2001;**25**:275-286

Section 2

Insecticides

Chapter 2

Defence against Oxidative Stress and Insecticides in *Musca domestica*

Tan Yong Hao, Siti Nasuha Hamzah and Zazali Alias

Abstract

This review is looking at the way *Musca domestica* defends itself against harmful molecules. One of the most notable enemies is against oxidative stress. Over the years there were reports that indicated the development of resistance on range of pesticides that are used against the flies. Researches have demonstrated that there are several functional protein molecules which contribute directly or indirectly as a response to oxidative stress and resistance against insecticides. As currently, the whole genome sequencing of the organisms has enabled future study to be conducted in evaluating the behaviour of the targeted protein/enzyme in response to oxidative stress and intake of insecticides in the flies.

Keywords: *Musca domestica*, oxidative stress, insecticide resistance

1. Introduction

An estimated 150,000 of species of Diptera have been described [1], and houseflies (*Musca domestica*) are one of the wonderfully evolved organism. A notorious vector, houseflies are associated with more than 100 pathogens [2], and resistance towards insecticides of houseflies have been reported all over the world. According to Scott et al., *Musca domestica* is suitable as a model organism for resistance studies and development of new insecticides. The knowledge on cellular metabolism in recent years has been expanded to understand the metabolic aspect of oxidative stress. In *Musca domestica* alone, a few families of proteins have been more or less associated with oxidative stress response: glutathione S-transferases (GST) [3–5], superoxide dismutase (SOD) [6] and glutathione peroxidases [7–9].

2. *M. domestica* response towards insecticide

Naturally houseflies' main ecosystem role is to decompose and recycle organic material. Houseflies are synanthropic insect in urban areas where high densities of human waste are their food source [10, 11]. It has been known to be vectors of various diseases of over 30 bacteria, protozoan, viruses and helminth eggs [12]. It also transfers viruses such as polioviruses [13] and *Coxackie* viruses, as well as numerous bacteria such as *Campylobacter jejuni*, *Helicobacter pylori* [14], *Salmonella* sp. [14], *Listeria* sp., *Yersinia pseudotuberculosis* [15], *Shigella* sp. [16], *Escherichia coli* [17], and *Vibrio* sp. [13]. Flies may also be vectors of protozoan flies such as *Giardia* and

Entamoeba [16] and eggs of several tapeworms [18]. In 2010, there were further proof on transmission of Newcastle disease virus (NDV—*Paramyxoviridae*), a highly infectious virus shed in the faeces in infectious birds [19] with *Musca domestica* as vector in both field and laboratory. More recently, *Musca domestica* were also reported to carry antibiotic-resistant bacteria such as methicillin-resistant *Staphylococcus* and ticarcilin-resistant *Pseudomonas* [20], which possess threat in hospitals and healthcare facilities [18]. Flies are causing 6 million cases of childhood blindness each year (http://www.who.int/topics/trachoma/en/). *Musca domestica* also create implications in economical ways, and costs of pesticides were estimated at more than US$200 million yearly in the United States [21] and US$1.6 million in 1998 [22].

The types of insecticides used to control houseflies on field are adulticides and larvicides (www.flycontrol.norvatis.com). Adulticides are carbamates (e.g. propoxur and methomyl), organophosphates (e.g. fenitrophon, azamethiphos and dimethoate), pyrethroids (e.g. cyfluthrin, deltamethrin and permethrin) and recently neonicotinoids (e.g. imidacloprid, thiamethoxam). Larvicides are insect growth regulators (IGRs) (e.g. triflumuron, diflubenzuron, cyromazine [23], and novaluron and juvenile hormone synthetic analogues (e.g. methoprene, fenoxycarb, pyriproxyfen [23] (www.flycontrol.novartis.com). Since the first case of DDT resistance is reported on the housefly [24], resistance of adult *Musca domestica* towards various insecticides in various sites (agricultural, wild and urban) is a fast-growing global issue. There has been an increasing resistance profile report from various places in the world.

In the United Kingdom, a resistance risk assessment done by [25] showed that although farmers claimed they had reduced using insecticides (a measure to reduce selective stress on field housefly strains), there was no sign of decrease of housefly resistance towards piperonyl butoxide synergized pyrethrins. Flies with high fenitrothion and dimethoate resistance were also discovered in Denmark [26]. In 1997, an increase in pyrethroid-resistant strains and widespread of azamethiphos-resistant strains in 21 different farms all over Denmark were confirmed [27]. In Argentina, a first insecticide survey was reported [28]. Several *Musca domestica* populations were found to be permethrin-, dichlorovinyl dimethyl phosphate (an organophosphate)- and cyromazine-resistant. In the neighbouring Brazil, [29] led a first evaluation of cyromazine resistance of houseflies in five different sites, and three out of the five sites indicated cyromazine resistance. There was a report suggesting the occurrence of insecticide tolerance in tsunami-hit villages in India [30]. With hygiene at minimum provision, immediate fly control was imposed by spraying 76% dichlorvos, and LD90 of adult housefly was 3.5–3.9 times higher than the flies from control sites. As in the United States, in a study tested against nine insecticides, the fly strains showed high resistances in tetrachlorvinphos, permethrin and cyfluthrin [31], while in southeastern Nebraska, houseflies are shown to be moderately resistant to permethrin yet extremely resistant to stirofos and methoxychlor [32]. Deltamethrin-resistant flies were discovered in urban garbage dump of cities of Beijing, Tianjin and Zhangjiakou [33].

In Malaysia, [34] resistance of housefly from a garbage dump, poultry farm and agricultural farm was evaluated. It was shown that garbage dump and poultry farm fly samples were more resistant than agricultural farm. It was also shown that two poultry farms in the state of Penang against malathion, propoxur and DDT, with resistance ratio, have been found with strong correlations against relative humidity, which is a first in field discovery [35]. However, on housefly larvae, resistance assessment has been relatively scarce with only a handful of feeding and toxicity tests done. A report on an increase in diflubenzuron resistance and new-found cyromazine resistant strain was also obtained [36]. A dip test-emergence test of *Musca domestica* third instar larvae on eucalyptol extracts has been done [37] with LD50 values of 118 mg/fly and 177 mg/fly on male and female flies, respectively.

3. Impact of oxidative stress-induced resistance

In an oxidative stress-induced insecticide resistance research, rats [38–40], humans [41], fresh water fish *Brycon cephalus* [42] and black tiger shrimp *Penaeus monodon* [43] have been used as models to investigate insecticide inflicted oxidative stress. Insecticides including pyrethroids [44, 45], organophosphates [46–48] and organochlorines [49] have known to be inducing oxidative stress. It was reported that there were changes in activities of the antioxidative enzymes such as superoxide dismutase, catalase, glutathione peroxidase and glutathione reductase and in GSH level changes both in the liver and erythrocyte homogenate [39]. Molecular resistances are consisted of target site resistance and metabolic-based resistance [50]. Yet, most of the works, as far as *Musca domestica* is concerned, have been more in top-down approach. While genome sequencing was still ongoing for that time being, specific gene family is identified and sequenced before getting into expression studies. With other fly species such as the dipteran *Drosophila melanogaster* and *Anopheles* genome as comparable reference database, it was also concluded three groups of gene superfamilies are involved in metabolic-based resistance [51], i.e. glutathione S-transferases, cytochrome P450 and acetylcholinesterase.

In cytochrome P450, [52] it was revealed that three P450 genes, CYP4D4v2, CYP4G2, and CYP 6A38, were up-regulated in response to permethrin treatment on permethrin-resistant ALHF strains. By using PCR technology, constant overexpression of CYP 6A1, CYP 6D1 and CYP 6D3 in neocotinoid-resistant strains in Denmark during thiomethoxam challenge was demonstrated [53]. CYP6D1 was also found to be implicating more than 5000-fold of cypermethrin resistance in Learn pyrethroid-resistant strain found in New York [54]. Significant increase in non-specific esterases and glutathione S-transferases activities were also evaluated [34]. A remarkable drop on GST activity has been reported on a DDT-resistant strain 698ab [27]. Point mutation was reported as the cause of insecticide sensitivity in the case of acetylcholinesterases (E.C 3.1.1.7) [55]. As far as metabolic-based resistance is concerned, there are still much more questions to be addressed. A study [31] stated that there is very little knowledge about the mechanism of the pyrethroid resistance (monooxygenase/CYP450), although pathways have been elucidated via genomic means. There was a significant correlation between kdr allele (i.e. genes reducing the sensitivity of the nervous system to pyrethroids) frequencies and the levels of knockdown resistance by deltamethrin via a PCR-based assay [33]. It was also demonstrated that a behavioural resistance might be playing a role in contributing such resistance and such traits are still being inherited in the field [25]. The upregulation mediated by changes to transacting factors reveals that these mechanisms were underlying in some cases of resistances of P450, GSTs, and acetylcholinesterases [56, 57].

3.1 Enzymatic removal of cellular hydrogen peroxide

It was suggested that aerobic organisms survive due to their evolved antioxidant capability [58]. Catalase (EC 1.11.1.6) was discovered in tobacco extracts [59]. Catalase detoxifies H_2O_2 into water and oxygen [60]. Catalase is one of the well-described enzymes, and it is a class of enzyme including the iron-heme enzyme, catalase-peroxidases and a small group of manganese enzymes [61]. Superoxide dismutase (EC 1.15.1.1) is a well-known enzyme against oxidative stress. SOD1, the first superoxide dismutase to be identified, uses free radical as a substrate [62]. A metalloenzyme, superoxide dismutase catalyses the dismutation of superoxide anion (O^{2-}) to hydrogen peroxide and oxygen, as the first defence line against oxidative stress [63]. They are also known to exhibit additional peroxidase activity when hydrogen peroxide level is at its large. It has been suggested that the removal of superoxide anion will reduce SOD alternate toxic behaviour [6]. Copper-zinc and manganese SODs scavenge and dismutate superoxide anion in mitochondrial electron transport systems. It was demonstrated that a manganese superoxide dismutase-deficient yeast thrived in hyperoxia conditions (95% oxygen, 5% carbon dioxide) under the removal of electron transport system [64]. A copper-zinc SOD1 in baker's yeast was characterized at the intermembrane space of mitochondria [65].

Glutathione peroxidase (EC 1.11.1.9) utilizes reduced glutathione (GSH) to decompose hydrogen peroxide [66–68]. This enzyme was discovered [66] and identified as selenocysteine enzymes at first [69], better known as GPx1. Later, more selenocysteines were identified such as GPxs-GPx2, GPx3, and GPx4 [70]. It was also found in mammals [68, 71]. Later, a catalytic cysteine residue on rat was discovered [72], known as GPx5, and followed by GPx6 [73] which is a selenocysteine proteins in humans but not in rats or mice [74]. Mammalians GPx7 and GPx8 were the last to be elucidated but have a low GPx activity [75].

Peroxiredoxins (EC 1.11.1.15) are another group of enzymes worth mentioning when discussing about oxidative stress in cellular organisms. Peroxiredoxins are a family of antioxidant enzymes [76]. Highly specific in reducing hydrogen peroxide [77], its cysteine residue makes up the active site of peroxiredoxins, which in turn are being oxidized into sulfenic acid and recycled back to thiol, via sulfiredoxins [78]. They also control cytokine-induced peroxide levels which, in turn, mediate signal transduction in mammalian cells [79].

4. Oxidative stress-related proteins in *Musca domestica*

There are several possible candidates of oxidative stress defence proteins. Those are superoxide dismutase, catalase, glutathione peroxidase, glutathione S-transferases, GSSG reductase, thiol transferases, gamma-glutamylcysteine synthetase, and glucose-6-phosphate dehydrogenase. Oxidative stress hypothesis is evident on aging and has always been raising questions from researchers. *Musca domestica* [80], *Drosophila melanogaster* [81, 82] and *Caenorhabditis elegans* [83] are made as model tested on hyperoxia conditions. Aging is resulted from oxidative damage from cellular macromolecules [81]. It was stated that the main prediction of this hypothesis is that the rate of aging cannot be slowed down without corresponding attenuation of oxidative damage/stress [84].

GST gene family and their isoforms have been discovered to participate in oxidative stress pathway. Overexpression and peroxidase activity of GSTs in peroxide treatment were observed [85]. Other than oxidative stress resistance, GSTs detoxify xenobiotics, protect from tissue damage, participate in Jun-kinase signaling pathway and act as non-catalytic carrier proteins (ligandins) in the intracellular transport of hydrophobic compounds [3–5]. Glutathione synthetase (GSHs) are responsible in the antioxidant defence as the dominant non-protein sulphhydryls in the cell [86] forming conjugates non-enzymatically or more by the catalysis and mediation of GSTs. H_2O_2 oxidizes thiolate group in cysteine residues (-S-) into thiols (-SOH), which is present in the exposing active site. Reaction against peroxidants is also energy-consuming due to the inhibition of oxidative phosphorylation [87] and deprives energy to maintain the recycling of NADPH during pentose phosphate pathway and glucose 6-phosphate dehydrogenase, making cells hyperglycaemic [88] and able to topple the condition of cell redox levels in levels of lactate/pyruvate ratio [89]. Most of the cases above were investigated towards organophosphates and pyrethroids. In cadmium ion treatment, concentration ranging from 0.2 to 5 mM in the medium, widely known to enhance reactive oxygen species in cell, increases the levels of superoxide dismutase [90]. Lowering the intake of selenium via diet increases the events of a peroxidative injury. The group further purified the selenium-independent glutathione peroxidase [8] and suggested this enzyme and the related pathways should be in the picture during the investigation of insect antioxidant defence system. There was no direct research work on peroxiredoxins with relation to houseflies, and its mechanisms and activities in vivo are not much of knowledge. However, it was discovered that there was no increase in catalase activity even though the diet of selenium in *Musca domestica* was lowered [7]. Another investigation [9] in houseflies revealed that the total inhibition of catalase also did not affect the survival of the flies, although slight increase in the level of SOD activity was observed.

5. Conclusion

Despite such remarkable immunity and rising insecticide tolerance exhibited by *Musca domestica*, and being such prominence as model for biochemistry and insect physiology, no genome project has been launched till 2009 [2]. More importantly, to the best of our knowledge, only a handful of *Musca domestica*-related proteomic work has been reported. However, in this last 5 years, there is an increasing interest unravelling the inner molecular workings of this insect. A genome project was launched [2], and a full genome of *Musca domestica* was successfully sequenced [91]. The sequenced genome is 691 MB, and some gene sequences notably 771 putative immune-related, 86 CYP450-related, and 33 glutathione S-transferase and 92 are predicted to have esterase activities. In comparison, this genome contained a plethora of shared and novel sequences than its *Drosophila* counterparts, supporting the fact of an exemplary ability of *Musca domestica* of associating closely with numerous amounts of pathogens living in a septic environment. Pioneering transcriptomic works have been done on *Musca domestica* larvae, by massive cDNA parallel pyrosequencing [92]. Thus with the help of recent advancement, a better insight on the mechanisms that are associated with oxidative stress and resistance against insecticides in *Musca domestica* is better understood.

Acknowledgements

The work has been funded by the University of Malaya Postgraduate Research Fund (PPP: PV091/2011A) and Ministry of Higher Education under the Fundamental Research Grant (FRGS: FP052-2014A).

Author details

Tan Yong Hao[1], Siti Nasuha Hamzah[2] and Zazali Alias[1,3]*

1 Faculty of Science, Institute of Biological Sciences, University of Malaya, Kuala Lumpur, Malaysia

2 School of Biological Sciences, Universiti Sains Malaysia, Pulau Pinang, Malaysia

3 University of Malaya Halal Research Centre (UMHRC), University of Malaya, Kuala Lumpur, Malaysia

*Address all correspondence to: alias@um.edu.my

References

[1] Thompson FC. Biosystematic Database of World Diptera. 2004. Retrieved from: www.diptera.org Trachoma. (n.d.). In: World Health Organization's website. Retrieved from: http://www.who.int/topics/trachoma/en/

[2] Scott JG, Liu N, Kristensen M, Clark AG. A case for sequencing the genome of *Musca domestica* (Diptera: Muscidae). Journal of Medical Entomology. 2009;**46**(2):175-182

[3] Ketterer B. Glutathione S-transferases and prevention of cellular free radical damage. Free Radical Research. 1998;**28**:647-668

[4] Strange RC, Jones PW, Fryer AA. Glutathione S-transferase: Genetics and role in toxicology. Toxicology Letters. 2000;**15**:357-363

[5] Yin Z, Ivanov VN, Habelhah H, Tew K, Ronai Z. Glutathione S-transferase elicits protection against H_2O_2-induced cell death via coordinated regulation of stress kinases. Cancer Research. 2000;**60**:4053-4057

[6] den Hartog GJM, Haenen GRMM, Vegt E, van der Vijgh WJF, Bast A. Superoxide dismutase: The balance between prevention and induction of oxidative damage. Chemico-Biological Interactions. 2003;**145**:33-39

[7] Simmons TW, Jamail IS, Lockshin RA. The effect of selenium deficiency on peroxidative injury in the house fly, Musca domestica: A role for glutathione peroxidase. FEBS Letters. 1987;**218**(2):251-254

[8] Simmons TW, Jamail IS, Lockshin RA. Selenium-independent glutathione peroxidase activity associated with glutathione S-transferase from the housefly. *Musca domestica. Comparative Biochemistry and Physiology Part B: Comparative Biochemistry*. 1989;**94**(2):323-332

[9] Allen RG, Farmer KJ, Sohal RS. Effect of catalase inactivation on levels of inorganic peroxides, superoxide dismutase, glutathione, oxygen consumption and life span in adult houseflies (*Musca domestica*). Biochemical Journal. 1983;**216**(2):503-506

[10] Dahlem G. House fly (*Musca domestica*). In: Resh V, Carde R, editors. Encyclopedia on Insects. 1st ed. Vol. 1. San Diego: CA Academic Press; 2003. pp. 532-534

[11] Marshall S. Insects: Their Natural History and Diversity. Buffalo, New York: Firefly Books Ltd; 2006

[12] Greenberg B, Kowalski JA, Klowden MJ. Factors affecting the transmission of salmonella by flies: Natural resistance to colonization and bacterial interference. Infection and Immunity. 1970;**2**:800-809

[13] Kettle DS. Muscidae (houseflies, stableflies). In: Kettle DS, editor. Medical and Veterinary Entomology. Wallingford: CAB International; 1990. pp. 223-240

[14] Grubel P, Hoffman JS, Chong FK, Burstein NA, Mepani C, Cave D. Vector potential of houseflies (Musca domestica) for Helicobacter pylori. Journal of Clinical Microbiology. 1997;**35**:1300-1303

[15] Zurek L, Denning SS, Schal C, Watson DW. Vector competence of *Musca domestica* (Diptera: Muscidae) for *Yersinia pseudotuberculosis*. Journal of Medical Entomology. 2001;**38**:333-335

[16] Ugbogu OC, Nwachukwu NC, Ogbuagu MN. Isolation of *Salmonella* and *Shigella* species from houseflies

(*Musca domestica* L.) in Utru, Nigeria. The African Journal of Biotechnology. 2006;**5**:1090-1091

[17] Szalanski AL, Owens CB, Mckay T, Steelman CD. Detection of *Campylobacter* and *Escherichia coli* O157:H7 from filth flies by polymerase chain reaction. Medical and Veterinary Entomology. 2004;**18**:241-246

[18] Graczyk TK, Knight R, Gilman RH, Cranfield MR. The role of non biting flies in the epidemiology of human infectious diseases. Microbes and Infection. 2001;**3**:231-235

[19] Barin A, Arabkhazaeli F, Rahbari S, Madani SA. The housefly, *Musca domestica*, as a possible mechanical vector of Newcastle disease virus in the laboratory and field. Medical and Veterinary Entomology. 2010;**24**:88-90

[20] Boulesteix G, Le Dantec P, Chevalier B, Dieng M, Niang B, Diatta B. Role of Musca domestica in the transmission of multiresistant bacteria in the centres of intensive care setting in a sub-Saharan Africa. Annales Francaises d'Anesthésie et de Réanimation. 2005;**24**:361-365

[21] Geden CJ, Arends JJ, Axtell RC, Barnard DR, Gaydon DM, Hickle LA, Hogstette JA, Jones WF, Mullens BA, Nolan Jr. MP, Petersen JJ, and Sheppard DC. Economic significance of poultry. In: Geden CJ, and Hogsette JA, editors, Research and extension needs for integrated pest management for arthropods of veterinary importance: Proceedings of a Workshop in Lincoln; 12-14 April 1994; Nebraska Gainesville, Florida: Center for Medical, Agricultural, and Veterinary Entomology USDA-ARS; pp. 1-328

[22] Crespo DC, Lecuona RE, Hogsette JA. Biological control: An important component in integrated management of *Musca domestica* (Diptera: Muscidae) in caged layer poultry houses in Buenos Aires. Biological Control. 1998;**13**:16-24

[23] Kočišová A, Petrovský M, Toporčák J, Novák P. The potential of some insect growth regulators in housefly (*Musca domestica*) control. Biologia. 2004;**59**:661-668

[24] Saccà GS. Estenza di *mosche domestiche* resistenti al DDT. Rivista di Parassitologia. 1947;**8**:127-128

[25] Learmount J, Chapman P, Macnicoll A. Impact of an insecticide resistance strategy for house fly (Diptera: Muscidae) control in intensive animal units in the United Kingdom. Journal of Economic Entomology. 2002;**95**(6):1245-1250

[26] Keiding J. The House Fly—Biology and Control. Geneva: World Health Organization; 1986

[27] Kristensen M. Glutathione S-transferase and insecticide resistance in laboratory strains and field populations of *Musca domestica*. Journal of Economic Entomology. 2005;**98**(4):1341-1348

[28] Acevedo GR, Zapater M, Toloza AC. Insecticide resistance of house fly, *Musca domestica* (L.) from Argentina. Parasitology Research. 2009;**105**:489-493

[29] Pinto MC, Prado APD. Resistance of *Musca domestica* L. populations to cyromazine (insect growth regulator) in Brazil. Memórias do Instituto Oswaldo Cruz. 2001;**96**(5):729-732

[30] Srinivasan R, Jambulingam P, Gunasekaran K, Boopathidoss P. Tolerance of housefly, *Musca domestica* L. (Diptera: Muscidae) to dichlorvos (76% EC) an insecticide used for fly control in the tsunami-hit coastal villages of southern India. Acta Tropica. 2008;**105**:187-190

[31] Scott JG, Alefantis TG, Kaufman PE, Rutz DA. Insecticide resistance in house flies from caged-layer poultry

facilities. Pest Management Science. 2000;**56**:147-153

[32] Marçon PCRG, Thomas GD, Siegfried BD, Campbell JB, Skoda SR. Resistance status of house flies (Diptera: Muscidae) from Southeastern Nebraska beef cattle feedlots to selected insecticides. Journal of Economic Entomology. 2003;**96**(3):1016-1020

[33] Cao XM, Song FL, Zhao TY, Dong YD, Sun CHX, Lui BL. Survey of deltamethrin resistance in house flies (*Musca domestica*) from urban garbage dumps in northern China. Environmental Entomology. 2006;**35**(1):1-9

[34] Nazni WA, Ursula MP, Lee HL, Sa'diyah I. Susceptibility of Musca domestica L. (Diptera: Muscidae) from various breeding sites to commonly used insecticides. Journal of Vector Ecology. 1999;**24**(1):54-60

[35] Bong LJ, Zairi J. Temporal fluctuations of insecticides resistance in *Musca domestica* Linn. (Diptera: Muscidae) in Malaysia. Tropical Biomedicine. 2010;**27**(2):317-325

[36] Kristensen M, Jespersen JB. Larvicide resistance in *Musca domestica* (Diptera: Muscidae) populations in Denmark and establishment of resistant laboratory strains. Journal of Economic Entomology. 2003;**96**(4):1300-1306

[37] Sukontason KL, Boonchu N, Sukontason K, Choochote W. Effects of eucalyptol on housefly (Diptera: Muscidae) and blow fly (Diptera: Calliphoridae). Revista do Instituto de Medicina Tropical de São Paulo. 2004;**46**(2):97-101

[38] Lukaszewicz-Hussain A, Moniuszko-Jakoniuk J. Activities of superoxide dismutase and catalase in erythrocytes and concentration of malondialdehyde in serum of rats intoxicated with chlorfenvinphos in low doses. Polish Journal of Environmental Studies. 1999;**8**:234-236

[39] Lukaszewicz-Hussain A, Moniuszko-Jakoniuk J. Organophosphate insecticide chlorfenvinphos affects enzymatic and nonenzymatic antioxidants in erythrocytes and serum of rats. Polish Journal of Environmental Studies. 2003;**12**:417-423

[40] El-Demerdash FM. Lipid peroxidation, oxidative stress and acetylcholinesterase in rat brain exposed to organophosphate and pyrethroid insecticides. Food and Chemical Toxicology. 2011;**49**:1346-1352

[41] Ranjbar A, Solhi H, Mashayekhi FJ, Susanabdi A, Rezaie A, Abdollahi M. Oxidative stress in acute human poisoning with organophosphorus insecticides; a case control study. Environmental Toxicology and Pharmacology. 2005;**20**:88-91

[42] Monteiro DA, Rantin FT, Kalinin AL. The effects of selenium on oxidative stress biomarkers in the freshwater characid fish matrinxã, *Brycon cephalus* (Günther, 1869) exposed to organophosphate insecticide Folisuper 600 BR® (methyl parathion). Comparative Biochemistry and Physiology (Part C). 2009;**149**:40-49

[43] Dorts J, Silvestrea F, Huynh TT, Tybergheina AE, Nguyen TP, Kestemonta P. Oxidative stress, protein carbonylation and heat shock proteins in the black tiger shrimp, Penaeus monodon, following exposure to endosulfan and deltamethrin. Environmental Toxicology and Pharmacology. 2009;**28**:302-310

[44] Giray B, Gürbay A, Hincal F. Cypermethrin-induced oxidative stress in rat brain and liver is prevented by vitamin E or allopurinol. Toxicology Letters. 2001;**118**:139-146

[45] Gupta A, Nigam D, Gupta A, Shukla GS, Agarwal AK. Effect of pyrethroid-based liquid mosquito repellent inhalation on the blood-brain barrier function and oxidative damage in selected organs of developing rats. Journal of Applied Toxicology. 1999;**19**:67-72

[46] Akhgari M, Abdollahi M, Kebryaeezadeh A, Hosseini R, Sabzevari O. Biochemical evidence for free radical-induced lipid peroxidation as a mechanism for subchronic toxicity of malathion in blood and liver of rats. Human and Experimental Toxicology. 2003;**22**:205-211

[47] Fortunato JJ, Agostinho FR, Réus GZ, Petronilho FC, Dal-Pizzol F, Quevedo J. Lipid peroxidative damage on malathion exposure in rats. Neurotoxicity Research. 2006;**9**(1):23-28

[48] Lukaszewicz-Hussain A. Role of oxidative stress in organophosphate insecticide toxicity—Short review. Pesticide Biochemistry and Physiology. 2010;**98**(2):145-150

[49] Koner BC, Banerjee BD, Ray A. Organochlorine pesticide-induced oxidative stress and immune suppression in rats. Indian Journal of Experimental Biology. 1998;**36**(4):395-398

[50] Perry T, Batterham P, Daborn PJ. The biology of insecticidal activity and resistance. Insect Biochemistry and Molecular Biology. 2011;**41**:411-422

[51] Ranson H, Claudinos C, Ortelli F, Abgrall C, Hemingway J, Sharakova MV, et al. Evolution of supergene families associated with insecticide resistance. Science. 2002;**298**:179-181

[52] Zhu F, Li T, Zhang L, Liu N. Co-up-regulation of three P450 genes in response to permethrin exposure in permethrin resistant house flies, Musca domestica. BMC Physiology. 2008;**8**:18

[53] Kristensen M, Spencer AG, Jespersen JB. The status and development of insecticide resistance in Danish populations of the housefly *Musca domestica* L. Pest Management Science. 2001;**57**(1):82-89

[54] Scott JG, Georghiou GP. Rapid development of high-level permethrin resistance in a field-collected strain of housefly (Diptera: Muscidae) under laboratory selection. Journal of Economic Entomology. 1985;**78**:316-319

[55] Fournier D. Mutations of acetylcholinesterase which confer insecticide resistance in insect populations. Chemico-Biological Interactions. 2001;**157-158**:257-261

[56] Feyereisen R. Insect P450 enzymes. Annual Review of Entomology. 1999;**44**:507-533

[57] Hemingway J. The molecular basis of two contrasting metabolic mechanisms of insecticide resistance. Insect Biochemistry and Molecular Biology. 2000;**30**:1009-1015

[58] Halliwell B, Gutteridge JMC. Free Radicals in Biology and Medicine. 4th ed. Oxford: Clarendon Press; 2006

[59] Loew O. A new enzyme of general occurrence in organisms. Science. 1900;**11**(279):701-702

[60] Keilin D, Hartree EF. On the mechanism of the decomposition of hydrogen peroxide by catalase. Proceedings of the Royal Society of London - Series B: Biological Sciences. 1938;**124**(837):397-405

[61] Nicholls P. Classical catalase: Ancient and modern. Archives of Biochemistry and Biophysics. 2012;**525**:95-101

[62] McCord JM, Fridovich I. Superoxide dismutase: An enzymic function for erythrocuprein. The

Journal of Biological Chemistry. 1969;**224**(22):6049-6055

[63] Fridovich I. Superoxide radical and superoxide dismutases. Annual Review of Biochemistry. 1995;**64**:97-112

[64] Guidot DM, McCord JM, Wright RM, Repine JE. Absence of electron transport (Rho 0 state) restores growth of a manganese-superoxide dismutase-deficient Saccharomyces cerevisiae in hyperoxia. Evidence for electron transport as a major source of superoxide generation in vivo. The Journal of Biological Chemistry. 1993;**268**:26699-26703

[65] Sturtz LA, Diekert K, Jensen LT, Lill R, Culotta VC. A fraction of yeast Cu, Zn-superoxide dismutase and its metallochaperone, CCS, localize to the intermembrane space of mitochondria: A physiological role for SOD1 in guarding against mitochondrial oxidative damage. The Journal of Biological Chemistry. 2001;**276**:38084-38089

[66] Mills GC. Hemoglobin catabolism. Glutathione peroxidase, an erythrocyte enzyme which protects hemoglobin from oxidative breakdown. The Journal of Biological Chemistry. 1957;**266**:20752-20760

[67] Forstrom JW, Zakowski JJ, Tappel AL. Identification of the catalytic site of rat liver glutathione peroxidase as selenocysteine. Biochemistry. 1978;**17**(13):2639-2644

[68] Ursini F, Maiorino M, Gregolin C. The selenoenzyme phospholipid hydroperoxide glutathione peroxidase. Biochimica et Biophysica Acta. 1985;**839**:62-70

[69] Chance B, Sies H, Boveris A. Hydroperoxide metabolism in mammalian organs. Physiological Reviews. 1979;**59**:527-605

[70] Chu FF, Doroshow JH, Esworthy RS. Expression characterization, and tissue-distribution of a new cellular selenium-dependent glutathione peroxidase, GSHPx-GI. The Journal of Biological Chemistry. 1993;**268**:2571-2576

[71] Takahashi K, Avissar N, Whitin J, Cohen H. Purification and characterization of human plasma glutathione peroxidase: A selenoglycoprotein distinct from the known cellular enzyme. Archives of Biochemistry and Biophysics. 1987;**256**:677-686

[72] Ghyselinck NB, Dufaure JP. A mouse cDNA sequence for epididymal androgen regulated proteins related to glutathione peroxidase. Nucleic Acids Research. 1990;**18**:7144

[73] Dear TN, Campbell K, Rabbitts TH. Molecular cloning of putative odorant-binding and odorant metabolizing proteins. Biochemistry. 1991;**30**:10376-10382

[74] Kryukov GV, Castellano S, Novoselov SV, Lobanov AV, Zehtab O, Guigó R, et al. Characterization of mammalian selenoproteomes. Science. 2003;**300**:1439-1443

[75] Brigelius-Flohe R, Maiorino M. Glutathione peroxidases. Biochimica et Biophysica Acta. 1830;**2012**:3289-3303

[76] Kim K, Kim IH, Lee KY, Rhee SG, Stadtman ER. The isolation and purification of a specific "protector" protein which inhibits enzyme inactivation by a thiol/Fe(III)/O_2 mixed-function oxidation system. The Journal of Biological Chemistry. 1988;**263**:4704-4711

[77] Peskin AV, Low FM, Paton LN, Maghzal GJ, Hampton MB, Winterbourn CC. The high reactivity of peroxiredoxin 2 with H_2O_2 is not

reflected in its reaction with other oxidants and thiol reagents. The Journal of Biological Chemistry. 2007;**282**(16):11885-11892

[78] Woo HA, Jeong W, Chang TS, Park KJ, Park SJ, Yang JS, et al. Reduction of cysteine sulfinic acid by sulfiredoxin is specific to 2-cys peroxiredoxins. The Journal of Biological Chemistry. 2005;**280**:3125-3128

[79] Hofmann B, Hecht HJ, Flohé L. Peroxiredoxins. Biological Chemistry. 2002;**383**(3-4):347

[80] Yan LJ, Levine RL, Sohal R. Effects of aging and hyperoxia on oxidative damage to cytochrome c in the housefly, *Musca domestica*. Free Radical Biology and Medicine. 2000;**29**(1):90-97

[81] Magwere T, West M, Riyahi K, Murphy MP, Smith RAJ, Partridge L. The effects of exogenous antioxidants on lifespan and oxidative stress resistance in *Drosophila melanogaster*. Mechanisms of Ageing and Development. 2005;**127**:356-370

[82] Rebrin I, Sohal R. Comparison between the effects of aging and hyperoxia on glutathione redox state and protein mixed disulfides in *Drosophila melanogaster*. Mechanisms of Ageing and Development. 2006;**127**:869-874

[83] Leiers B, Kampkötter A, Grevelding CG, Link CD, Johnson TE, Henkle Dührsen K. A stress-responsive glutathione S-transferase confers resistance to oxidative stress in Caenorhabditis elegans. Free Radical Biology and Medicine. 2003;**34**(11):1405-1415

[84] Sohal RS, Weindruch R. Oxidative stress, caloric restriction, and aging. Science. 1996;**273**:59-63

[85] Veal EA, Toone WM, Jones N, Morgan BA. Distinct roles for glutathione S-transferases in the oxidative stress response in *Schizosaccharomyces pombe*. The Journal of Biological Chemistry. 2002;**277**(38):35523-35531

[86] Ketterer B. The role of nonenzymatic reactions of glutathione in xenobiotic mechanism. Drug Metabolism Reviews. 1982;**13**:161-187

[87] Milatovic D, Gupta RC, Aschner M. Anticholinesterase toxicity, oxidative stress. The Scientific World Journal. 2006;**6**:295-310

[88] Rahimi R, Abdollahi M. A review on the mechanisms involved in hyperglycemia induced by organophosphorus pesticides. Pesticide Biochemistry and Physiology. 2007;**88**:115-121

[89] Lukaszewicz-Hussain A, Moniuszko-Jakoniuk J. Procesy glikolityczne w wa̧trobie szczurawzatruciu chlorfenwinfosem. Medycyna Pracy. 1997;**48**:580-583

[90] Dabas A, Nagpure NS, Kumar R, Kushwaha B, Kumar P, Lakra WS. Assessment of tissue-specific effect of cadmium on antioxidant defense system and lipid peroxidation in freshwater murrel, Channa punctatus. Fish Physiology and Biochemistry. 2012;**38**(2):469-482

[91] Scott JG, Warren WC, Beukeborn LW, Bopp D, Clark AG, Giers SD, et al. Genome of the house fly, *Musca domestica* L., a global vector of diseases with adaptations to a septic environment. Genome Biology. 2014;**15**:466

[92] Liu F, Tang T, Sun L, Priya TAJ. Transcriptomic analysis of the housefly (*Musca domestica*) larva using massively parallel pyrosequencing. Molecular Biology Reports. 2012;**39**:1927-1934

Chapter 3

Neonicotinoid Insecticides: A Threat to Pollinators

Muhammad Imran

Abstract

Pollination is the fundamental requirement for healthier fruit set. More than 90% of flowering plant species in the hot and humid regions required pollination. Many plants species required animal pollination. Among these animals, insects play a vital role in pollination, and among the major insect pollinators, hymenopterans, honeybees, and bumblebees are regarded as the best pollinators of the crops around the world. Declining population of these important pollinators day by day is a major threat, and this declining is due to a variety of stressors. Among these possible reasons including environmental conditions, parasites, predators, malnutrition, and diseases, many researchers pointed finger at pesticides as playing a major role especially neonicotinoid. Neonicotinoids move in the environment and can be found throughout the areas where they are not applied. Neonicotinoids can drift offsite directly exposing bees and contaminate nontargeted areas when applied as spray. During plant uptake, neonicotinoid spreads through plant tissues and disrupts the physiology of pollen eaters, nectar feeders, and the insects that feed upon plant tissues. Therefore, the use of neonicotinoid is the major reason for the decline of bees in the world. So it is requested to all farmers and researchers to please find ways to kill pests not pollinators.

Keywords: pollinators, honeybees, bumblebees, insecticides, Neonicotinoid

1. Economic importance of pollinators

Consideration on sustainable growth generally agrees that environment still harbors much kind of living things that are potentially and unswervingly significant to mankind. Their lucrative utilize is now pending for the discovery of their worth or the formulation of how they should be propagated. There are about 25,000 species of bees [1] recognized in the world and only few play an important role in pollination producing fruits and seeds. Most of the world wide plant species depend upon animal pollination for their fertilization [2]. Among animal pollinators (any animal which transfer pollen between plants enabling fertilization and sexual reproduction from anther of male flower part to the stigma of female flower) insects provide better service of pollination [3]. Insect pollinators include bees (honey bees, bumblebees and solitary bees), flies (Carrion flies, flesh flies and hover flies), pollen wasp, ants, mosquitoes, beetles, butterflies and moths [4, 5].

Among these major insect pollinators; hymenopterans, honey bees and bumblebees are regarded as the best pollinators of the crops around the world. It has been introduced globally due to its economic importance of honey production (honey bee) and pollination of the crops [5]. Bees are known to pollinate among 71 most familiar crops out of hundred plant species that accounts for 90% of world's food supply [6]. However, honey bees and bumblebees are the principal pollinators of the crops and it has been used successfully as pollinators in crop systems around the world [7, 8].

Many fields of current agriculture hang on pollinators. In each pollination season, these important pollinators mostly honey bees, bumblebees and native bees bring billions of U.S dollars in economic value. In several esteems, they play as a key role in the world economy [9]. But it is very important to know the real value of these important little creatures. About $230 and $580 billion U.S. dollars' worth of annual worldwide food production depend on the direct influence of these important pollinators [10].

Managed bees (domesticated bees by the beekeeper) are the greatest regarded pollinators in relations of agricultural economics. These pollinators (honeybees and bumblebees) can deliver pollination to almost any crop. Almond crop is entirely reliant upon honey bee pollination. Without these pollinators, yield for many fruit crops including watermelon, squash, blueberries and other fruits would be greatly reduced [11, 12]. According to the statistic presented by USDA, a honeybee colony value 100 times more to the public than to the beekeeper it mean that the value they deliver extends well beyond their actual price. Bee's pollination has aided make vegetables, nuts and fruits more accessible to consumers. There are many others species of insects called as wild species like leaf cutter bees, mason bees, alfalfa bees are not documented for their input to current agriculture. But these pollinating insects provide supplement to managed bees colonies but also pollinate some crops more professionally than their managed bees. Throughout blooming season honeybees and other native insects partner to deliver pollination for many crops. Although the economic values of their pollination is much less than managed bees, but the role of wild bees is important [11].

2. Ecosystem essentials

Preserving our indigenous flora, including wild for example bluebells, poppies, cornflowers and, along with trees, also be contingent on pollinator populations. This is much closer relationship between the declining of pollinator's population and the plant they pollinate and this relationship goes parallel throughout the world [13, 14]. It is estimated that in Europe and UK about 76% of plants that are pollinated by or called as liked by bumblebees have declined in recent decades. Pollinator's population declines spell bad news for previously declining wildflowers, which are pollinated mostly by insects and among them one fourth are endangered. In short wildlife also depends on these important pollinators, declining of wild flora means declining of wildlife including birds their shelter. Even though the insects themselves provide a significant link in the food chain as prey for other insects, birds and other animals that feed on insect [15, 16].

3. Current declines in pollinator populations

To maintain the plant genetic diversity pollination is very important for plant reproduction [12]. Due to its important role in agriculture many scientist worked on population dynamics of these important pollinators. There are many reason behind the decreasing population of these important pollinators such as bats, beetles, flies birds and bees, the main reasons behind this are habitat destruction [17, 18] and the introduction chemicals sprayed on crops in form of pesticides [11, 19]. Monitoring programs of NASS led by the USDA have documented the decline in managed bee's population since 1947, making them the most important example of pollinator decline in North America [11, 20]. Reasons behind the decline of these important pollinators including managed and wild bees are of mites that feed on honeybee larvae and adult body making them weak, pathogens, use of antibiotics to control these pathogen and pesticides [21–23]. Among these all factors pesticides paly vial role for the declining of population. A huge amount of these pesticides are sprayed on crops for the control of insect pest that damage crops, and bees are non-target organism on these sprayed crops. When bees visit on these sprayed crops to collect pollen and nectar become contaminated. Among these pesticides many are neuro-toxic in nature such as parathion, diazinon, and carbaryl play vial role in population decline [21].

However, the population of honeybee is declining day by day due to intemperate uses of pesticides [8]. Generally the bees are exposed to these pesticides; which are either used to control the parasitic mites and the pathogens attacked in the hives or to control the diseases and pest in the crops on which the bees are visited for pollen and nectar [21]. The experiments conducted in Europe and the United States found the miscellaneous range of pesticides on healthy and unhealthy bee's colonies along with their pollen, honey and bee waxes [12]. One possible cause of distressing bee mortality is the use of very active systemic insecticide called neonicotinoids [19].

4. Neonicotinoid, a real threat to pollinators

Neonicotinoids; systemic insecticides, easily soluble in water but slowly break in the environment. These insecticides are absorbed by the plants through roots system and become the part of plant. The photo-degradation, half-life of neonicotinoids is about 30 $\pm$ 4 days when exposed to sunlight [24]. It is highly toxic to insects as compare to mammals and birds because they are unable to cross the blood-brain barrier due to the lack of a charged nitrogen atom and the uncharged molecule can penetrate the insect blood–brain barrier [25]. It is derived from nicotine, which is accountable for bees decline and are highly selective neuro-active insecticides [26].

Neonicotinoids were introduced into the market in 1990 [27]. This new class of insecticide is neurotoxic, includes imidacloprid, thiamethoxam, dinotefuran, nitenpyram, acetamiprid, thiacloprid and clothianidin [28]. The first commercial neonicotinoid was imidacloprid meanwhile clothianidin and thiamethoxam were the first two introduced insecticides in early 2000s in the market [27]. Neonicotinoids are systemic poisons acquire by plants through their root system and they may endured in the plants for weeks to months and mostly depends on the abiotic conditions and application rate [29, 30]. Neonicotinoids are used to protect a variety of vegetables, fruits, and major crops like corn, cotton, potato, rice, etc. against sucking insects like aphids, whiteflies, thrips, leaf- and plant hoppers [31]. In Pakistan, these insecticides are recommended for the control of sucking pests of cotton, as they are most effective against thrips, jassid, and whitefly [32, 33].

The insecticides having the neonicotinoid compounds were applied on 140 different crops in more than 120 countries around the world. The excessive use of the neonicotinoids has been reported as the major factor in declining of both domestic and wild bees. Neonicotinoids are broad spectrum insecticides and are moderately to highly effective and toxic to bees that depends upon the presence of active ingredient in the insecticides [34]. Neonicotinoids are mainly used in seed and soil treatment and sometimes they also directly applied to plant foliage [27]. Many of the neonicotinoids are highly toxic to the insect pollinators and also to the honey bees. It changes the behavior that results in the behavioral disturbances, orientation difficulties and impairment of social activities [35–41].

Neonicotinoids also affects the CNS (central nervous system) of the insects as it binds agonistically to the post-synaptic nicotinic acetylcholine receptors that results in the spontaneous discharge of nerve impulses and eventual failure of the neuron to propagate any signal [42]. The neonicotinoids and their metabolites have the capability to persistent in the soil and aquatic sediments [43] and their persistence at shallow depths could increase the chances of aquatic life and other wildlife including honey bees could get exposed to the insecticide [44].

The neonicotinoids are considered to be most effective insecticide other than organophosphates and carbamates [45]. Imidacloprid is the most widely used insecticide and has drawn more attention on the health of bees than other neonicotinoids. More than 400 products of this insecticides accounting for about 15th of the globally insecticide marketed [46]. Honeybees are exposed to neonicotinoids in different ways from ingestion, contact and inhalation [47]. The pollen foragers which are different from the nectar foragers; they do not consume pollen by itself but it brings to the hives to consumed for the nurse bees and larvae hence the nurse bees and larvae exposed to neonicotinoids and their metabolites [48].

The forager bees used honey from their hive before they leave for foraging. It depends upon the distance that it will travel from their hive to foraging field, the forager bees have to consume more or less amount of nectar or honey from their hive for energy and foraging. Therefore the foragers may ingest more or less amount of residues of neonicotinoids [49]. The colony become contaminated when the worker bees come into contact with pollen or nectar contaminated with neonicotinoid and transport them to the hive, where they are normally observed in honey and bee bread [50, 51]. Bee hives made up of trees treated with neonicotinoids could have residues which may cause trouble for bees [52]. Oral route of neonicotinoid uptake is highest in forager honeybees, winter honeybees and larvae [53, 54]. Serious pests of citrus in Pakistan and other Asian countries are mostly control by using various classes of neonicotinoids. The foraging bees visiting citrus flowers get exposed to the residues of neonicotinoids which are responsible for damaging their physiology [55].

Neonicotinoids increased worker mortality and queenlessness over time. The toxicity of the neonicotinoids increases when it encountered with fungicides. In corn growing areas, the health of honey bees are reduced when are exposed to neonicotinoids in the field [56]. The irretrievable and cumulative damage to central nervous system of insects is often caused by neonicotinoid insecticides. There is no safe level of neonicotinoids and even only a very minute quantity of these systemic poisons could have long lasting drastic effects [57]. The activities of the acetyl cholinesterase is increased by the thiamethoxam at each developmental stages of the insects and the activities of glutathione-S- transferase and carboxyl esterase para increases at the pupal stages and reduced the survival of larvae and pupa that results in the decreasing of percentage emergence of honeybees [58]. The effects of thiamethoxam cause the reduction of forager bees returning to the hive [59]. When honey bees are exposed to a sub-lethal doses of imidacloprid and clothianidin that results in the reduction of foraging activities as well as longer foraging flights [60]. The bees become detract when it became exposed to nonlethal doses to thiamethoxam and causes high mortality at levels that may collapse the colony.

Among distinctive behaviors of honey bees, foraging is one of idiosyncratic behavior of the *Apis mellifera*. This type of behavior is like an association between the bee colonies and the ambient environment [59].

5. Conclusion

After World War II, we started using pesticides on a large scale, and this became necessary because of the monocultures that put out a feast for crop pest. Recently, researchers from Penn State University has started looking at the pesticides residue in the loads of pollen that bees carry home as food, and they have found that every batch of pollen that honeybee collects has at least six detectable pesticides in it, and this includes every class of insecticides, herbicides, fungicides and even inert and unlabeled ingredients that are part of the pesticides formulation that can be more toxic than the active ingredient. One of these classes of insecticides, the neonicotinoids is making headlines around the world right now you have probably heard about it. This is the new class of insecticides, it move through the plant so that a crop pest, a leaf eating insect would take a bite of plant and get a lethal dose and die. In most agricultural settings, on most of our farms it's only the seed that's coated with insecticides and so a smaller concentration move through the plant and gets into the pollen and nectar, and if a bee consumes this lower dose either nothing happens or the bee becomes intoxicated and disoriented and she may not find her way to home.

5.1 Strategies to conserve the pollinators

Every one of us needs to behave a little bit more like a bee society, and insect society, where each of our individual actions can contribute to grand solution and emergent property. So let the small act of planting flowers and keeping them free of pesticides be the driver to large scale change. Please find the ways to kill pest not bees.

Author details

Muhammad Imran
Department of Entomology, University of the Poonch, Rawalakot, Azad Jammu and Kashmir, Pakistan

*Address all correspondence to: imran.bees@gmail.com

References

[1] Michener CD. The Bees of the World. Baltimore: John Hopkins, London Press; 2000. p. 873

[2] Fulton M, Hodges SA. Floral isolation between *Aquilegia formosa* and *A. pubescens*. Proceedings of the Royal Society of London. 1999;**266**:2247-2252

[3] Hodges SA, Whittall JB, Fulton M, Yang JY. Genetics of floral traits influencing reproductive isolation between Aquilegia *formosa* and *A. pubescens*. American Naturalist. 2002; **159**:51-60

[4] Moreti ACCC, Silva ECA, Alves MLTMF. Observações sobre a polinização entomófila da cultura da soja (Glycine max Merril). Boletim da Indústria Animal. 1998;**55**:91-94

[5] Winfree R, Williams NM, Gaines H, Ascher JS, Kremen C. Wild bee pollinators provide the majority of crop visitation across land-use gradients in New Jersey and Pennsylvania, USA. Journal of Applied Ecology. 2008;**45**: 793-802

[6] Gallaia N, Sallesc JM, Setteled J, Vaissierea BE. Economic valuation of the vulnerability of world agriculture confronted with pollinator decline. Ecological Economics. 2009;**68**:810-821

[7] Artz DR, Hsu CI, Nault B. Influence of honey bee, *Apis mellifera*, hives and field size on foraging activity of native bee species in pumpkin fields. Environmental Entomology. 2011;**40**: 1144-1158

[8] Morse RA, Calderone NW. The value of honey bee as pollinator of U.S. crops. Bee Culture. 2000;**128**:1-15

[9] Kremen C, Williams NM, Aizen MA, Herren BG, LeBuhn G, Minckley R, et al. Pollination and other ecosystem services produced by mobile organisms: A conceptual framework for the effects of land-use change. Ecology Letters. 2007;**10**:299-314

[10] Vanbergen AJ, Insect Pollinators Initiative. Threats to an ecosystem service: Pressures on pollinators. Frontiers in Ecology and the Environment. 2013;**11**:251-259

[11] Berenbaum M, Bernhardt P, Buchmann S, Calderone NW, et al. Status of Pollinators in North America. Washington, D.C: The National Academies Press; 2007. p. 322

[12] Delaplane KS, Mayer DF. Crop Pollination by Bees. New York: CABI Publishing; 2000. p. 344

[13] Morales CL, Aizen MA. Does invasion of exotic plants promote invasion of exotic flower visitors? A case study from the temperate forests of the southern Andes. Biological Invasions. 2002;**4**:87-100

[14] Parker IM. Pollinator limitation of *Cytisus scoparius* (scotch broom), an invasive exotic shrub. Ecology. 1997;**78**: 1457-1470

[15] Levine JM, Vilà M, Antonio CM, Dukes JS, Grigulis K, Lavorel S. Mechanisms underlying the impacts of exotic plant invasions. Proceedings of the Royal Society of London. Series B, Biological Sciences. 2003;**270**:775-781

[16] Richardson DM. Plant invasion ecology dispatches from the front line. Divers. Diversity and Distributions. 2004;**10**:315-319

[17] Stephen WP. Alfalfa pollination in Manitoba. Journal of Economic Entomology. 1995;**48**:543-548

[18] Aizen MA, Feinsinger P. Habitat fragmentation, native insect pollinators, and feral honey bees in argentine 'Chaco

Serrano'. Ecological Applications. 1994;**4**: 378-392

[19] Kevan PG. Forest application of the insecticide fenitrothion and its effect on wild bee pollinators (Hymenoptera: Apoidea) of lowbush blueberries (Vaccinium spp.) in Southern New Brunswick, Canada. Biological Conservation. 1975;7:301-309

[20] (NASS) National Agricultural Statistics Service. Honey. Washington, D.C.: United States Department of Agriculture; 1976-2008. Available from: http://www.nass.usda.gov, Dec 2008

[21] Johansen CA, Mayer DF. Pollinator Protection: A Bee and Pesticide Handbook. Cheshire, CT: Wicwas Press; 1990. p. 322

[22] Morse RA, Flottum K. Honey Bee Pests, Predators, and Diseases. Medina, OH: A.I. Root Company; 1997. p. 454

[23] Wilson WT, Pettis JS, Henderson CE, Morse RA. Tracheal mites. In: Morse RA, Flottum K, editors. Honey Bee Pests, Predators, and Diseases. 3rd ed. Medina, OH: A.I. Root Company; 1997. pp. 253-278

[24] Cressey D. Europe debates risk to bees. Nature. 2013;**496**:408-409

[25] Izuru Y. Nicotine to nicotinoids. In: Yamamoto I, Casida J, editors. Nicotinoid Insecticides and the Nicotinic Acetylcholine Receptor. Springer-Verlag; 1999. pp. 3-27

[26] Cresswell JEE, Desneux N, Vangelsdrop D. Dietary traces of neonicotinoid pesticides are a cause of population declines in honeybees. Pest Management Science. 2012;**68**:819-827

[27] Tomizawa M, Casida J. Neonicotinoid insecticide toxicology: Mechanisms of selective action. Annual Review of Pharmacology and Toxicology. 2005;**54**:247-268

[28] Jeschke JM. Across islands and continents, mammals are more successful invaders than birds. Diversity and Distributions. 2008;**14**:913-916

[29] Bromilow RH, Chamberlain K, Evans AA. Physiochemical aspects of phloem transclucation of herbicides. Weed Science. 1990;**38**:305-314

[30] Laurent FM, Rathahao E. Distribution of imidacloprid in sunflowers (*Helianthus annuus*) following seed treatment. Journal of Agricultural Food Chemistry. 2003;**51**: 8005-8010

[31] Elbert A, Haas M, Springer B, Thielertf W, Nauen R. Applied aspects of neonicotinoid uses in crop protection. Pest Management Science. 2008;**64**: 1099-1105

[32] Bethke AJ, Redak RA. Effect of imidacloprid on the silver leaf whitefly, *Bemisia argentifolii* bellows and Perring (Homoptera: Aleyrodidae), and whitefly parasitism. Annals of Applied Biology. 2008;**130**:397-407

[33] Lopez JR, Jr D, Fritz BK, Latheef MA, Lan Y, Martin DE, et al. Evaluation of toxicity of selected insecticides against thrips on cotton in laboratory bioassays. Journal of Cotton Science. 2008;**12**:188-194

[34] Jeschke P, Nauen R, Schindler M, Elbert A. Overview of the status and global strategy for neonicotinoids. Journal of Agricultural Food Chemistry. 2011;**59**:2897-2908

[35] Guez D, Suchail S, Gauthier M, Maleszka R, Belzunces LP. Contrasting effects of imidacloprid on habituation in 7- and 8- day - old honeybees (*Apis mellifera*). Neurobiology Learning Memory. 2001;**76**:183-191

[36] Bortolotti L, Montanari R, Marcelino J, Medrzycki P, Aini S, Porrini C. Effects of sub-lethal

imidacloprid doses on the homing rate and foraging activity of honey bees. Bulletin of Insectology. 2003;**56**:63-67

[37] Medrzycki P, Montanari R, Bortolotti L, Sabatini AG, Maini S, Porrini C. Effects of Imidacloprid administered in sub-lethal doses on honey bees' behaviour. Laboratory tests. In: Proceedings of the 8th International Symposium "Hazards of Pesticides to Bees", September 4–6, 2002, Bologna, Italy (Eds). Bulletin of Insectology. Vol. 56. 2003. pp. 59-62

[38] Decourtye A, Armengaud C, Renou M, Devillers J, Cluzeau S, Gauthier M, et al. Imidacloprid impairs memory and brain metabolism in the honeybee (*Apis mellifera*). Pesticide Biochemistry and Physiology. 2004;**78**: 83-92

[39] Desneux N, Decourtye A, Delpuech JM. The sublethal effects of pesticides on beneficial arthropods. Annual Review of Entomology. 2007;**52**: 81-106

[40] Hassani AKEI, Dacher M, Gary V, Lambin M, Gauthier M, Armengaud C. Effects of sublethal doses of acetamiprid and thiamethoxam on the behavior of the honeybee (*Apis mellifera*). Archives Environmental Contamination Toxicology. 2008;**54**:653-666

[41] Maini S, Medrzycki P, Porrini C. The puzzle of honey bee losses: A brief review. Bulletin of Insectology. 2010;**63**: 153-160

[42] Matsuda K, Sattelle DB. Mechanism of selective actions of neonicotinoids on insect acetylcholine receptors. In: Clark JM, Ohkawa H, editors. New Discoveries in Agrochemicals: American Chemical Society Symposium Series. Oxford, UK: Oxford University Press; 2005. pp. 172-183

[43] Doering J, Maus C, Schoening R. Residues of imidacloprid in blossom samples of rhododendron sp after soil treatment in the field application. Environmental Chemistry. 2004;**62**: 483-494

[44] Lu Z, Jonathan K, Challis WCS. Quantum yields for direct photolysis of neonicotinoid insecticides in water: Implications for exposure to non-target aquatic organisms. Environmental Science and Technology. 2015;**2**:188-192

[45] Bonmatin JM, Marchand PA, Charvet R, Moineau I, Bengsch ER, Colin ME. Quantification of Imidacloprid uptake in maize crop. Journal of Agricultural and Food Chemistry. 2005;**53**:5336-5341

[46] Chen M, Lin T, Collins E, Lu MC. Simultaneous determination of residues in pollen and high fructose corn syrup from eight neonicotinoid insecticides by liquid chromatography-tandem mass spectrometry. Analytical and Bio analytical Chemistry. 2013;**405**: 9251-9264

[47] Girolami KM, Kahng SW, Hilker KA, Girolami PA. Differential reinforcement of high rate behavior to increase the pace of self-feeding. Behavioral Interventions. 2009;**24**:17-22

[48] Rortais A, Amold G, Halm MP, Briens TF. Modes of honey bee exposure to systemic insecticides: Estimated amounts of contaminated pollen and nectar consumed by different categories of bees in France. Apidologie. 2005;**36**: 71-83

[49] Maxim L, Slujis VD. Seed dressing systemic insecticide and honeybees. European Environment Agency. 2013; **376**:1-17

[50] Blacquiere T, Smagghe G, Van Gestel CA, Mommaerts V. Neonicotinoids in bees: A review on concentrations, side-effects and risk assessment. Ecotoxicology. 2012;**21**: 973-992

[51] Genersch E, Ohe VDW, Kaatz H, Schroeder A, Otten C, Buechler R, et al. The German bee monitoring project: A long term study to understand periodically high winter losses of honey bee colonies. Apidologie. 2010;**41**:332-352

[52] Beekman M, Ratnieks FLW. Long range foraging by honey bees *Apis mellifera*. Functional Ecology. 2000;**14**: 490-496

[53] Chauzat MP, Faucon JP, Martel AC, Lachaize CN, Aubert M, et al. Journal of Economical Entomology. 2006;**99**: 253-262

[54] Reetz JE, Zühlke S, Spiteller M, Wallner K. Neonicotinoid insecticides translocated in guttated droplets of seedtreated maize and wheat: A threat to honeybees? Apidologie. 2011;**42**: 596-606

[55] Khan AA, Afzal M, Raza AM, Khan AM, Iqbal J, Tahir HM, et al. Toxicity of botanicals and selective insecticides to Asian citrus psylla, *Diaphorina citri* K.(Homoptera: Psyllidae) in laboratory conditions. Jokull Journal. 2013;**63**:52-72

[56] Tsvetkov N, Robert OS, Sood K, Patel HS, Malena DA, Gajiwala PH, et al. Chronic exposure to neonicotinoids reduces honey bee health near corn crops. Science. 2017;**356**:1395-1397

[57] Tennekes HA. The significance of the Druckrey-Küpfmüller equation for risk assessment–the toxicity of neonicotinoid insecticides to arthropods is reinforced by exposure time. Toxicology. 2010;**276**:1-4

[58] Tavares DA, Dussaubat C, Kretzschmar A, Carvalho SM, Zacarin ECMS, Malaspina O, et al. Exposure of larvae to thiamethoxam affects the survival and physiology of the honey bee at post-embryonic stages. Environmental Pollution, Elsevier. 2017; **229**:386-393

[59] Henry M, Beguin M, Requier F, Rollin O, Odoux JF, Aupinel P, et al. A common pesticide decreases foraging success and survival in honey bees. Science. 2012;**336**:348-350

[60] Schneider S, Eisenhardt D, Rademacher E. Sublethal effects of oxalic acid on *Apis mellifera* (Hymenoptera: Apidae): Changes in behaviour and longevity. Apidologie. 2012;**43**:218-225

Section 3

Integrated Pest Management Strategies

Chapter 4

Management of *Spodoptera litura* (Fab.) in Green Gram (*Vigna radiata* L.) through Entomo-Pathogenic Nematode

Shakti Singh Bhati

Abstract

Green gram is most important legume crop and richest source of 24% easily digestible protein. The green gram is attacked by number of insect pests but *Spodoptera litura* is more serious pest. The uses of entomopathogenic nematodes (EPN) as a biological control agent of insect pests are more effective. EPNs have been found effective for the management of tobacco caterpillar and are used as bio insecticides against a number of lepidopteran pests. The mass multiplication of *Steinernema carpocapsae* can be done on rice moth (*Corcyra cephalonica*), greater wax moth (*Galleria mellonella*), gram pod borer (*Helicoverpa armigera*) and tobacco caterpillar (*Spodoptera litura*). Infectivity of entomopathogenic nematode, *S. carpocapsae* against tobacco caterpillar was studied and observation was recorded after every day up to 10 days with different inoculum levels *viz.*, 10,000, 15,000 and 20,000 IJs/plant of *S. carpocapsae*. The experimental results revealed that maximum 82.50% mortality of *S. litura* was observed at inoculum level 20,000 IJs/plant of *S. carpocapsae* after 9th day of inoculation followed by 75.00% mortality at inoculum level 15,000 IJs/plant. While, minimum 67.50% mortality was recorded at inoculum level 10,000 IJs/plant. Therefore, it was concluded that the mortality of insect larvae increased with an increase in the inoculum levels and period of exposure.

Keywords: green gram, infectivity, mass multiplication, *Spodoptera litura*, *Steinernema carpocapsae*

1. Introduction

Green gram (*Vigna radiata*) also known as mung bean, is native to India and Central Asia. The food legumes were grown by farmers since millennia providing nutritionally balanced food to the people of India [1] and many other countries in the world. Pulses occupy a unique position in economy of our nation being the major source of proteins. The major pulse crops that have been domesticated and are under cultivation include, green gram, black gram, chickpea, cowpea, pigeon pea, horse gram, lentil, moth bean, and pea.

Green gram is an important source of easily digestible high quality protein for vegetarians. It contains 24% protein, 0.326% phosphorus, 0.0073% iron, 0.00039%

carotene, 0.0021% of niacin [2]. Researchers has pointed out that plant protection remains a most neglected aspect in pulse cultivation; further stating that only 5–6% of the growers adopt plant protection measures in only 1.5% of the total area under this crop. The green gram is attacked by number of insect pests *viz. Helicoverpa armigera*, *Spodoptera litura*, *Maruca vitrata*, *Etiella zinckenella*, *Mylabris phalerata*. They cause significant damage to green gram including foliage and pods. The losses caused to green gram come to about 20%.

Spodoptera litura (Lepidoptera: Noctuidae) is a serious polyphagous pest of several cultivated crops and has attained global importance. The losses caused by *S. litura* on mung bean is much more severe as this pest has been reported to cause skelatalization of leaves in early stage and severe defoliation in later stage thus reducing the photosynthetic capacity of plants. Tobacco caterpillar (*Spodoptera litura*) has a wide host range of more than 120 host plants including crops (green gram, tobacco, soybean, castor, maize, sorghum, groundnut, linseed and mustard), vegetables (tomato, okra, brinjal and cucurbits) weeds and ornamental plants and the losses caused to these crops may range from 20 to 30% [3]. The caterpillars may eat entire leaves, and even flowers and fruits. The caterpillar burrows into the soil several centimeters and pupates without a cocoon. The pupal stage lasts either a few weeks or several months, depending upon time of year. The average life cycle is completed in about 25 days.

Realizing the role of these pests as limiting factor in agricultural productivity, several methods were developed and incorporated in to management program of the economically important pest. Out of these, use of insecticides could initially catch up to the growers because of their ready availability, ability to suppress pest's populations quickly and increasing productivity. Widespread development of resistance to chemical insecticides including the widely used pyrethroids has been reported in *S. litura* [4]. In addition to the development of resistance in pests, indiscriminate use of pesticides has grossly poisoned almost each and every component of the biosphere, including resurgence of pests and reduction of natural enemies in agro ecosystems, allowing rapid rebound of target and minor pests.

Use of insecticides although found effective however, looking into the adverse effect of chemical insecticides, several bioagents have been tried time to time to manage this pest but none of them could give desirable results.

Biological control of pests using entomopathogenic nematodes (EPNs) may prove to be an ideal alternative to other bioagent earlier used they have long term effect, without any harmful effect on non-target organisms. EPNs are potential agents as they serve as vectors of bacteria, achieve a quick kill of target insect pests, have broader host range, highly virulent, possess chemoreceptor's and can be cultured easily *in vitro* and *vivo*. EPNs can be easily applied using standard application equipment and are compatible with many chemical pesticides. The EPNs of the families *Steinernematidae* and *Heterorhabditidae* are potentially useful for biological control in agriculture systems [5]. The infective juveniles (IJs) of these families are free living, non-feeding and have the ability to search out their hosts. They have the potential for long-term establishment in soil through recycling of infected insects larvae. The importance of entomopathogenic nematode as a key component for the management of pests.

2. Mass multiplication of *Steinernema carpocapsae* on different hosts

Mass multiplication of *Steinernema carpocapsae* was done on rice moth (*Corcyra cephalonica*), greater wax moth (*Galleria mellonella*), gram pod borer (*Helicoverpa armigera*) and tobacco caterpillar (*Spodoptera litura*). The infective juveniles of *S. carpocapsae* were released @ 100 IJs/larvae into the petri plate having 4th instar larvae of different insect hosts allowing them to enter into the insect body. Harvesting of EPN's population was done after 10 days of inoculation using white trap method up to 5 days.

Results have indicated that, on the basis of per mg. body weight of cadaver maximum 572.00 IJs of *S. carpocapsae* were produced on *G. mellonella*, followed by 568.00 IJs and 554.00 IJs on *S. litura* and *H. armigera* respectively. Whereas, minimum 542.00 IJs on *C. cephalonica*. Therefore, *G. mellonella* was the most suitable host for mass production of *S. carpocapsae* (**Table 1**).

2.1 Rice moth (*Corcyra cephalonica*)

The data on yield of IJs presented in **Table 2** showed that maximum 60212.0 IJs of *S. carpocapsae* were produced on large sized larvae (14–16 mm) with mean body weight of 134 mg/larvae followed by 48320.0 IJs from medium sized larvae (10–12 mm) and 39635.0 IJs from small sized larvae (6–8 mm).

2.2 Greater wax moth (*Galleria mellonella*)

The data on yield of IJs presented in **Table 1** showed that maximum 100240.0 IJs of *S. carpocapsae* were produced on large sized larvae (18–20 mm) with mean body weight of 202 mg/larvae followed by 66036.0 IJs from medium sized larvae (13–15 mm) and 49252.0 IJs from small sized larvae (10–12 mm).

S. no.	Size of larvae (mm)	Mean weight (mg/larvae)	IJs harvested/larvae	IJs/mg body weight of cadaver
1.	Small (10–12)	86	49252.00	572.75
2.	Medium (13–15)	131	66036.00	504.25
3.	Large (18–20)	202	100240.00	496.25
	SEm±	3.543	662.457	8.504
	CD (5%)	11.334	2119.316	27.205
	CV (%)	5.07	1.84	3.24

Inoculum level = 100 IJs/larvae, replication = 4 times.

Table 1.
Yield of Steinernema carpocapsae from the larvae of greater wax moth (Galleria mellonella).

S. no.	Size of larvae (mm)	Mean weight (mg/larvae)	IJs harvested/ larvae	IJs/mg body weight of cadaver
1.	Small (6–8)	73	39635.00	542.94
2.	Medium (10–12)	90	48320.00	538.50
3.	Large (14–16)	134	60212.00	449.50
	SEm±	2.465	1163.214	4.976
	CD (5%)	7.886	3721.324	15.919
	CV (%)	4.98	4.72	1.95

Inoculum level = 100 IJs/larvae, replication = 4 times.

Table 2.
Yield of Steinernema carpocapsae from the larvae of rice moth (Corcyra cephalonica).

2.3 Gram pod borer (*Helicoverpa armigera*)

The data on yield of IJs presented in **Table 3** showed that maximum 115362.0 IJs of *S. carpocapsae* were produced on large sized larvae (30–32 mm) with mean body weight of 274 mg/larvae followed by 106070.0 IJs from medium sized larvae (25–27 mm) and 113590.0 IJs from small sized larvae (20–22 mm) (**Figures 1** and **2**).

S. no.	Size of larvae (mm)	Mean weight (mg/larvae)	IJs harvested/ larvae	IJs/mg body weight of cadaver
1.	Small (20–22)	205	113590.00	554.25
2.	Medium (25–27)	236	106070.00	449.25
3.	Large (30–32)	274	115362.00	421.00
	SEm±	3.976	1795.069	7.535
	CD (5%)	12.719	5742.737	24.105
	CV (%)	3.34	3.21	3.17

Inoculum level = 100 IJs/larvae, replication = 4 times.

Table 3.
Yield of Steinernema carpocapsae from the larvae of gram pod borer (Helicoverpa armigera).

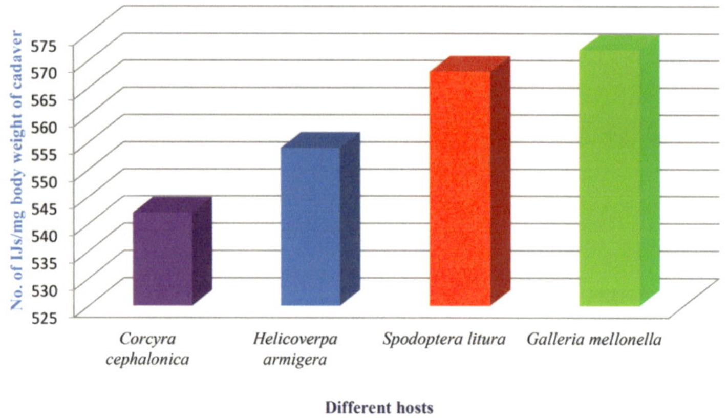

Figure 1.
Mass multiplication of Steinernema carpocapsae on different hosts.

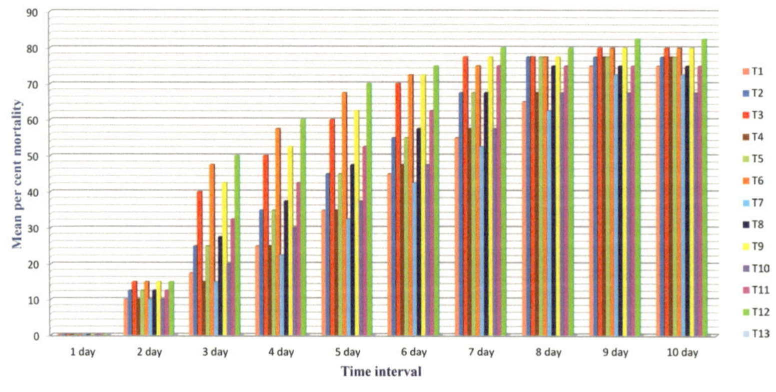

Figure 2.
Infectivity of Steinernema carpocapsae recovered from different hosts (a) Corcyra cephalonica, (b) Galleria mellonella, (c) Helicoverpa armigera and (d) Spodoptera litura *against* Spodoptera litura *infecting green gram.*

2.4 Tobacco caterpillar (*Spodoptera litura*)

The data on yield of IJs presented in **Table 4** showed that maximum 201280.0 IJs of *S. carpocapsae* were produced on large sized larvae (26–28 mm) with mean body weight of 430 mg/larvae followed by 200900.0 IJs from medium sized larvae (22–24 mm) and 193140.0 IJs from small sized larvae (18–20 mm) (**Figures 3** and **4**).

S. no.	Size of larvae (mm)	Mean weight (mg/larvae)	IJs harvested/ larvae	IJs/mg body weight of cadaver
1.	Small (18–20)	340	193140.00	568.00
2.	Medium (22–24)	400	200900.00	502.00
3.	Large (26–28)	430	201280.00	468.25
	SEm±	8.808	4406.542	11.159
	CD (5%)	28.179	14097.29	35.698
	CV (%)	4.52	4.44	4.35

Inoculum level = 100 IJs/larvae, replication = 4 times.

Table 4.
*Yield of Steinernema carpocapsae from the larvae of tobacco caterpillar (*Spodoptera litura*).*

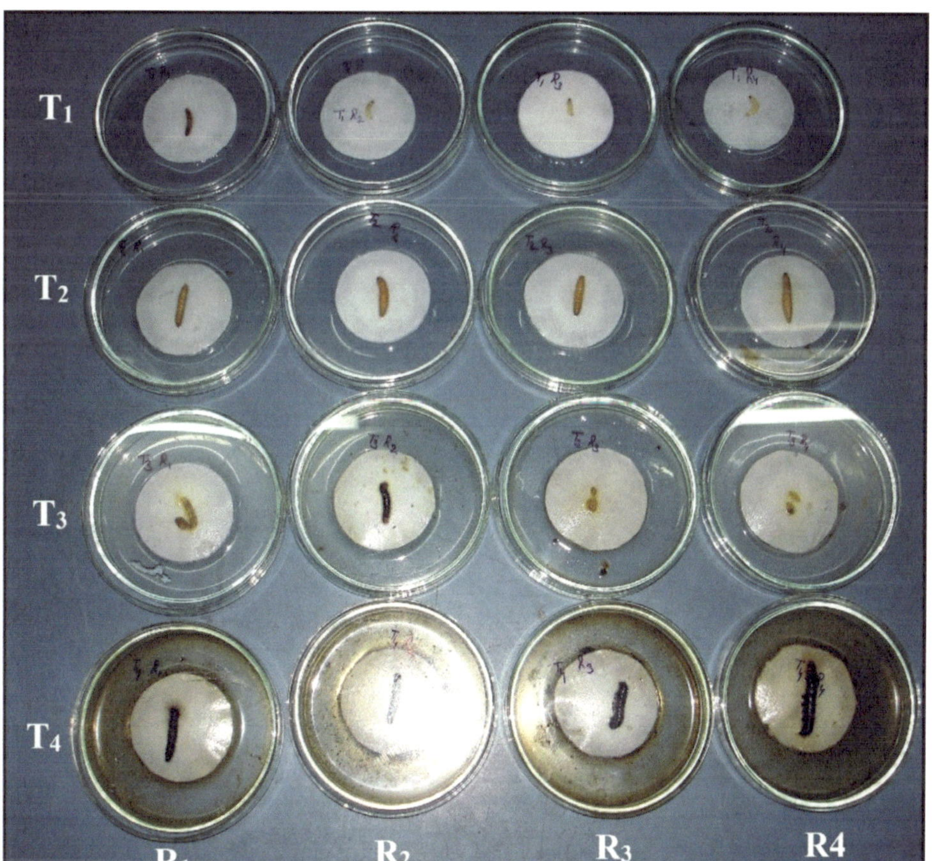

Figure 3.
Mass multiplication of Steinernema carpocapsae on different hosts.

Figure 4.
Infectivity of Steinernema carpocapsae recovered from different hosts against Spodoptera litura *infecting green gram.*

3. Infectivity of *Steinernema carpocapsae* recovered from different hosts against *Spodoptera litura* infecting green gram

Experiment was conducted to find out the infectivity of *S. carpocapsae* recovered from different natural hosts *viz. Corcyra cephalonica*, *Galleria mellonella*, *Helicoverpa armigera* and *Spodoptera litura* at different inoculum levels 10,000, 15,000 and 20,000. The mean percent mortality was recorded after every day up to 10 days.

3.1 After 1st day

The experimental results presented in **Table 5** revealed that there was no mortality of insect larvae, by inoculating IJs recovered from natural hosts *viz. C. cephalonica*, *G. mellonella*, *H. armigera* and *S. litura*.

3.2 After 2nd day

Results showed that 15.00, 12.50 and 10.00% mortality of *S. litura* was achieved at inoculum levels 20,000, 15,000 and 10,000 IJs/plant respectively, with populations recovered from *C. cephalonica*, *G. mellonella*, *H. armigera* and *S. litura*.

S. no.	Treatments	Mean percent mortality at different intervals									
		1 day	2 day	3 day	4 day	5 day	6 day	7 day	8 day	9 day	10 day
1.	$S_1 D_1$	0.00	10.00 (18.43)	17.50 (24.16)	25.00 (29.89)	35.00 (36.22)	45.00 (42.12)	55.00 (47.88)	65.00 (53.78)	75.00 (60.11)	75.00 (60.11)
2.	$S_1 D_2$	0.00	12.50 (20.47)	25.00 (29.89)	35.00 (36.22)	45.00 (42.12)	55.00 (47.88)	67.50 (55.28)	77.50 (61.77)	77.50 (61.77)	77.50 (61.77)
3.	$S_1 D_3$	0.00	15.00 (22.50)	40.00 (39.17)	50.00 (45.00)	60.00 (50.83)	70.00 (56.95)	77.50 (61.77)	77.50 (61.77)	80.00 (63.43)	80.00 (63.43)
4.	$S_2 D_1$	0.00	10.00 (18.43)	15.00 (22.50)	25.00 (29.89)	35.00 (36.12)	47.50 (43.56)	57.50 (49.33)	67.50 (55.28)	77.50 (61.77)	77.50 (61.77)
5.	$S_2 D_2$	0.00	12.50 (20.47)	25.00 (29.89)	35.00 (36.22)	45.00 (42.12)	55.00 (47.88)	67.50 (55.28)	77.50 (61.77)	77.50 (61.77)	77.50 (61.77)
6.	$S_2 D_3$	0.00	15.00 (22.50)	47.50 (43.56)	57.50 (49.33)	67.50 (55.28)	72.50 (58.45)	75.00 (60.11)	77.50 (61.77)	80.00 (63.43)	80.00 (63.43)
7.	$S_3 D_1$	0.00	10.00 (18.43)	15.00 (22.50)	22.50 (28.23)	32.50 (34.72)	42.50 (40.67)	52.50 (46.44)	62.50 (52.27)	72.50 (58.45)	72.50 (58.45)
8.	$S_3 D_2$	0.00	12.50 (20.47)	27.50 (31.39)	37.50 (37.66)	47.50 (43.56)	57.50 (49.39)	67.50 (55.28)	75.00 (60.11)	75.00 (60.11)	75.00 (60.11)
9.	$S_3 D_3$	0.00	15.00 (22.50)	42.50 (40.61)	52.50 (46.44)	62.50 (52.34)	72.50 (58.61)	77.50 (61.77)	77.50 (61.77)	80.00 (63.43)	80.00 (63.43)
10.	$S_4 D_1$	0.00	10.00 (18.43)	20.00 (26.19)	30.00 (33.05)	37.50 (37.73)	47.50 (43.56)	57.50 (49.33)	67.50 (55.28)	67.50 (55.28)	67.50 (55.28)
11.	$S_4 D_2$	0.00	12.50 (20.47)	32.50 (37.42)	42.50 (40.67)	52.50 (46.44)	62.50 (52.27)	75.00 (60.11)	75.00 (60.11)	75.00 (60.11)	75.00 (60.11)
12.	$S_4 D_3$	0.00	15.00 (22.50)	50.00 (45.00)	60.00 (50.77)	70.00 (56.79)	75.00 (60.11)	80.00 (63.43)	80.00 (63.43)	82.50 (65.47)	82.50 (65.47)
13.	Control	0.00	0.00	0.00	0.00	0.00	0.00	0.00	0.00	0.00	0.00

S. no.	Treatments	Mean percent mortality at different intervals									
		1 day	2 day	3 day	4 day	5 day	6 day	7 day	8 day	9 day	10 day
	SEm ±	0.00	1.703	2.217	1.817	1.707	1.860	1.443	1.486	1.445	1.445
	CD (5%)	0.00	4.872	6.342	5.197	4.882	5.322	4.128	4.251	4.134	4.134
	CV (%)	0.00	18.03	14.80	10.19	8.30	8.04	5.63	5.45	5.11	5.11

Replication = 4 times, (10 larvae/pot).
EPN population produced on different host: S_1 = Corcyra cephalonica; S_2 = Galleria mellonella; S_3 = Helicoverpa armigera; S_4 = Spodoptera litura. *Doses: D_1 = 10,000 IJs/plant; D_2 = 15,000 IJs/plant; D_3 = 20,000 IJs/plant.*

Table 5.
Infectivity of Steinernema carpocapsae recovered from different hosts against Spodoptera litura *infecting green gram.*

3.3 After 3rd day

Data pertaining to mean percent mortality of *S. litura* presented in **Table 5** revealed that maximum 50.00% mortality of *S. litura* was observed at an inoculums level of 20,000 IJs/plant of *S. carpocapsae* obtained from *S. litura*, followed by 47.50% mortality at 20,000 IJs/plant produced on *G. mellonella*. Whereas, minimum 15.00% mortality at 10,000 IJs/plant recovered from both *G. mellonella*, and *H. armigera*.

3.4 After 4th day

Results showed in **Table 5** revealed that maximum 60.00% mortality of *S. litura* was recorded at an inoculum level of 20,000 IJs/plant of *S. carpocapsae* recovered from *S. litura*, followed by 57.50% mortality at 20,000 IJs/plant produced on *G. mellonella*. Whereas, minimum 22.50% mortality at 10,000 IJs/plant obtained from *H. armigera*.

3.5 After 5th day

Maximum 70.00% mortality of *S. litura* was recorded at an inoculum level of 20,000 IJs/plant of *S. carpocapsae* obtained from *S. litura*, followed by 67.50% mortality at 20,000 IJs/plant produced on *G. mellonella*, whereas, minimum 32.50% mortality at 10,000 IJs/plant recovered from *H. armigera*.

3.6 After 6th day

Data pertaining to mean percent mortality of *S. litura* presented in **Table 5** revealed that maximum 75.00% mortality of *S. litura* was recorded at an inoculum level of 20,000 IJs/plant of *S. carpocapsae* recovered from *S. litura*, followed by 72.50% mortality at 20,000 IJs/plant produced on *G. mellonella* as well as *H. armigera*. While, minimum 42.50% mortality at 10,000 IJs/plant obtained from *H. armigera*.

3.7 After 7th day

Maximum 80.00% mortality of *S. litura* was observed at an inoculum level of 20,000 IJs/plant of *S. carpocapsae* obtained from *S. litura*, followed by 77.50% mortality recorded at 20,000 IJs/plant produced on *C. cephalonica* and *H. armigera*, while minimum 52.50% mortality at 10,000 IJs/plant recovered from *H. armigera*.

3.8 After 8th day

Results showed in **Table 5** revealed that maximum 80.00% mortality of *S. litura* recorded at an inoculum level of 20,000 IJs/plant of *S. carpocapsae* obtained from *S. litura*, followed by 77.50% mortality at 20,000 IJs/plant produced on *H. armigera*, *C. cephalonica* and *G. mellonella* and at 15,000 IJs/plant recovered from *C. cephalonica* and *G. mellonella*. Whereas, minimum 62.50% mortality at 10,000 IJs/plant recovered from *H. armigera*.

3.9 After 9th day

Data pertaining to mean percent mortality of *S. litura* presented in **Table 5** revealed that maximum 82.50% mortality of *S. litura* was recorded at an inoculum level of 20,000 IJs/plant of *S. carpocapsae* obtained from *S. litura*, followed by

80.00% mortality at 20,000 IJs/plant recovered from *C. cephalonica*, *G. mellonella* and *H. armigera*. Whereas, minimum 72.50% mortality recorded at 10,000 IJs/plant recovered from *H. armigera*.

3.10 After 10th day

Maximum 82.50% mortality of *S. litura* was recorded at an inoculum level of 20,000 IJs/plant of *S. carpocapsae* recovered from *S. litura*, followed by 80.00% mortality was recorded at 20,000 IJs/plant produced on *C. cephalonica*, *G. mellonella* and *H. armigera*. Whereas, minimum 72.50% mortality at 10,000 IJs/plant obtained from *H. armigera*.

4. Conclusion

EPNs are excellent biocontrol agents for insect pests. When an EPN is used against a pest insect, it is critical to match the right nematode species against the target pest. Biotic agents including nematode pathogens, predators and other soil organisms, as well as abiotic factors such as ultraviolet radiation, soil moisture/relative humidity, temperature, etc., can affect EPN application efficacy. Recently, improvement of nematode formulation, application equipment or approaches, and strain improvement have been made to enhance EPN application efficacy. Additional research toward lowering product costs, increasing product availability, enhancing ease-of-use, and improving efficacy and carryover effect will stimulate the extensive use of EPNs in biocontrol. With these advances EPNs will serve to reduce chemical insecticide inputs and contribute to the stabilization of crop yields and the environment.

In this chapter, we studied about the effect of host on multiplication and temperature on infectivity of *S. carpocapsae* against *S. litura* on green gram. Studies on mass multiplication of *Steinernema carpocapsae* was done on rice moth (*Corcyra cephalonica*), greater wax moth (*Galleria mellonella*), gram pod borer (*Helicoverpa armigera*) and tobacco caterpillar (*Spodoptera litura*). Results have indicated that on the basis of per mg body weight of cadaver maximum, *S. carpocapsae* was obtained from *G. mellonella*, followed *S. litura* and *H. armigera* respectively, whereas minimum IJs recovered from *C. cephalonica*. Therefore, it was concluded that on the basis of per mg body weight of cadaver *G. mellonella* was the most suitable host for mass production of *S. carpocapsae*.

When we studied about infectivity of *S. carpocapsae* against tobacco caterpillar (*S. litura*) under pot condition on green gram with different inoculum levels with different population of *S. carpocapsae* produce on natural hosts, the experimental results revealed that maximum percent mortality of *S. litura* was observed at 20000 IJs of *S. carpocapsae* recovered from *S. litura* after 9th days followed by 20,000 IJs recovered from *C. cephalonica*, *G. mellonella* and *H. armigera*. While, minimum percent mortality was recorded at 10,000 IJs recovered from *H. armigera*.

Author details

Shakti Singh Bhati
Nematology, College of Agriculture Nagaur, Agriculture University, Jodhpur, Rajasthan, India

*Address all correspondence to: shakti.singh3880@gmail.com

References

[1] Nene YL. Indian pulses through the millennia. Asian Agri-history. 2006;**10**:179-202

[2] Acharya SS. Prices and Price Policy for Pulses and Cereals. Udaipur XVIII+439+A: Sukhadia University; 1985. pp. 135

[3] Bhargava MC, Choudhary RK, Jain PC. Genetic engineering of plants for insect resistance. In: Jain PC, Bhargava MC, editors. Entomology: Novel Approaches. New Delhi, India: New India Publishing; 2008. pp. 133-144

[4] Ahmad M, Arif MI, Ahmad M. Occurrence of insecticide resistance in field populations of *Spodoptera litura* (Lepidoptera: Noctuidae) in Pakistan. Crop Protection. 2007;**26**:809-817

[5] Gaugler R, Kaya HK. Entomopathogenic Nematodes in Biological Control. Boca Raton, Florida: CRC Press, Inc.; 1990. p. 365

Chapter 5

Integrated Pest Management of the Yam Chip Beetle *Dinoderus porcellus* Lesne (Coleoptera: Bostrichidae): Current Status and Future Prospects

Loko Yêyinou Laura Estelle, Toffa Dèca Mondoukpè Joelle and Orobiyi Azize

Abstract

In West Africa, *Dinoderus porcellus* Lesne (Coleoptera: Bostrichidae) is a pest that attacks and spoils stored yam chips. Despite this fact, very little attention has been given to this pest, which could destroy up to 65% of stocks. In order to prevent any damages, farmers are widely using chemical substances for fighting against this pest despite their negative impacts on human health and environment. This chapter aims at proposing a solution approach and discussing the development of an integrated pest management strategy. The solution approach includes storage bags, varietal resistance, botanicals, and biological control. Further research should be done on the use of hermetic bags, essential oils, entomopathogens, insect growth regulators, pheromones, and their combined effects in the *D. porcellus* control.

Keywords: biological control, botanicals, *Dioscorea* sp., storage bags, varietal resistance

1. Introduction

Yam (*Dioscorea* sp.) is a plant that produces edible tubers that contribute to food security and poverty reduction in West Africa. With a production of 67,312,076 tons in 2017, representing 96.3% of world production, West Africa is rising to first place in the production and consumption of yam (236 kcal) per capita per day [1]. Yam tubers that are used in the preparation of various culinary dishes in West Africa [2] are very rich in carbohydrates, vitamins, and minerals [3]. In addition to the important trade in yams in West Africa [4], it is anchored in the sociocultural life of the populations as evidenced by the many festivals organized for the release of new yams [5].

Unfortunately, because of their high water content [5], fresh yam tubers are highly perishable with postharvest losses of up to 85% [6]. An alternative to the perishability of yam tubers is the transformation into chips, stabilized product by peeling, precooking, and drying in the sun [7]. The manufacture of yam chips, known in Benin, Nigeria, and Togo, makes it possible to preserve the surplus of the tubers for use during the lean season [8]. However, the yam chips are severely attacked by insects that cause significant losses by reducing them to an inconspicuous powder within a few months [7]. Various insect pests such as *Dinoderus porcellus* Lesne, *Araecerus fasciculatus* DeGeer, *Dinoderus bifoveolatus* Wollaston, *Palorus subdepressus* Wollaston, *Tribolium castaneum* Herbst, *Rhyzopertha dominica* Fabricius, *Lasioderma serricorne* Fabricius, *Sitophilus zeamais* Motschulsky, *Cathartus quadricollis* Guérin-Méneville, *Gnatocerus maxillosus* Fabricius, *Cryptolestes pusillus* Schönherr, *Carpophilus binotatus* Murray, *Carpophilus dimidiatus* Fabricius, and *Psocoptera* spp. were found in stored yam chips [7, 9, 10]. However, *D. porcellus* remains by far the most abundant and most damaging storage pest to yam chips [9–12].

The use of synthetic chemical insecticide is the main method of control used by farmers to protect stored yam chips against *D. porcellus* [7, 10]. However, most of the insecticides used by farmers are not specific for stored yam chips protection, and their misuse has led to numerous cases of food poisoning and deaths of whole families registered [13, 14]. Faced with this deplorable situation, it is important to use alternative methods that respect the environment and the populations' health.

Several alternative methods of *D. porcellus* control have been tested in recent years such as the use of botanicals [15], varietal resistance [16], biological control [17, 18], and physical tools [19]. However, effective protection of stored yam chip from *D. porcellus* requires integration of all these control methods. In this chapter we review the various methods of *D. porcellus* control, highlight the methods to be explored in the future, and discuss an integrated control method against this pest.

2. General description about *Dinoderus porcellus*

2.1 Description

Beetles of the *Dinoderus* genus are the smallest of all bostrichids [20]. According to Schäfer et al. [21], they are characterized by a very short forehead with a fronto-clypeal suture very little distinct; their antennas are composed of 9–11 segments with the second segment shorter than the first segment. *Dinoderus* spp. are also characterized by the last abdominal segment visible and curved; the pronotum having a perforated posterior surface is bordered anteriorly by a row of teeth; the subapical carina is absent in elytra [21]. The adult of *D. porcellus* is reddish-brown with elytra black and hard, shiny, appearing glazed, almost glabrous on their dorsal part (**Figure 1**). *D. porcellus* is different from the other *Dinoderus* by a pronotum without fovea but with a posterolateral carina reaching the first row of marginal teeth and a dorsal punctuation of the elytra consisting of large perforations [21].

2.2 Biology

D. porcellus digs holes on the yam chips in which it reproduces. The female lays her eggs inside the dug holes, which hatches after 6–8 days [19]. It was noted that at a temperature below 10°C and above 40°C, no oviposition was observed in females of *D. porcellus* [19]. The larva development time is an average of 24 days, while the nymphal stage lasts an average of 5.25–6.50 days [19]. According to Nwana [22] the average development time of *D. porcellus* on yam chips is on average 35.9 days at room temperature.

2.3 Pest status

In West Africa, the beetle *D. porcellus* is a serious pest of stored chips of several roots and tubers such as cassava [21], yam [10], and cocoyam [22]. *D. porcellus* digs holes in the yam chips (**Figure 2**), drastically reducing their internal parts into powdery waste, which negatively affects their visual quality and decreases their commercial value. Under laboratory conditions, the weight loss due to this pest in

Figure 1.
Dinoderus porcellus adult (source: Schafer et al. [21]).

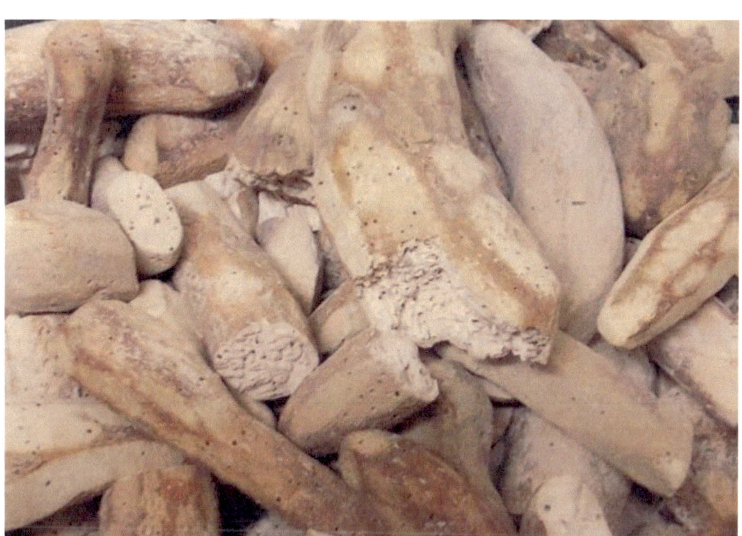

Figure 2.
Yam chips infested by D. porcellus.

4.5 months was estimated to 29.5% in newly dried yams and 39.2% in stock that had already been in store for 6 months [11]. In addition, when yam chips were infested with *D. porcellus* and stored for 3 months, the reconstituted thick paste (amala or télibo) was unsuitable for human consumption and not accepted by consumers [9].

3. Management of *Dinoderus porcellus* in yam chips

3.1 Physical methods

Physical control of stored product insects involves the manipulation of physical factors (temperature, relative humidity, atmospheric composition, etc.) to eliminate them or reduce their population to a tolerable level [23]. The control of stored product insect such as *D. porcellus* could be done by the use of heating or cooling of storage structures. Traditionally, infested yam chips are sun-dried by farmers [10]. Although sun-drying remains one of the oldest methods of control used in the protection of dried foods, it has several disadvantages such as exposure to waste and deterioration of vitamins [24]. In addition, this method is not very effective against *D. porcellus*, which feeds inside the chips so as not to be exposed to sunlight. Oni and Omoniyi studies [19] show that storage of yam chips at 20–30°C is optimal for the development and reproduction of *D. porcellus.* Lethal temperatures varied significantly with the species of yam used for making chips. In general, it is therefore recommended to store yam chips at temperatures below 20°C and above 35°C to control *D. porcellus* [19]. However, the manipulation of the temperature of storage structures requires infrastructures such as gas-tight containers, which are not accessible to smallholder farmers who are engaged in the processing of yam chips in West Africa. An alternative could be the use of triple-bagging consisting to seal dried food in a series of two heavy-grade polyethylene plastic bags which were expected to be as hermetic [25]. The use of the Purdue Improved Crop Storage (PICS) triple layer bag as an alternative to the use of the common polyethylene bags has been tested by Hell et al. [26] and have proven ineffective in protecting cassava chips against *Dinoderus* spp. and *Prostephanus truncatus*. Nevertheless, further research is needed to fill the knowledge gaps and provide adequate information needed to inform decision-maker for the use of PICS triple layer bag for yam chips protection against *D. porcellus*.

3.2 Botanical insecticides

Botanical insecticides are one of promising alternative to the use of chemical synthetic insecticides in pest control because of their minimal costs and ecological side effects [27]. For the protection of stored yam chips against insect pests, Vernier et al. [7] evaluated the level of protection provided by various biological products derived from neem (*Azadirachta indica* A. Juss.) and *Crotalaria sericea* L. compared with a synthetic pesticide, Sofagrain (1.5% deltamethrin +0.5% pirimiphos methyl). Among the tested organic products, oil, powdered leaves, and neem seeds gave the best level of protection in terms of reducing the damage caused by pests [7]. Eze et al. [28] had reported the potentials of ethanolic extracts of *A. indica* and *Ocimum gratissimum* L. to protect chips against insect pests. The insecticidal and repellent effects of the powders and extracts of three medicinal plants (*Bridelia ferruginea* Benth., *Blighia sapida* Koenig, and *Khaya senegalensis* (Desr.) A. Juss.) traditionally used in Benin by farmers for the protection of yam chips against *D. porcellus* were tested by Loko et al. [15]. The results showed that the leave powders of the three medicinal plants have strong repellent properties against *D. porcellus* and could be a source of novel repellent against this pest. The propanol extract of *B. ferruginea* at 5% proved to be a good fumigant against *D. porcellus*, with 88.89% of pest mortality at 160 μL/L air, while the acetone extract of *K. senegalensis* could be used in the development of a contact insecticide against *D. porcellus* because of having a low LC_{50} of 0.29 μL/insect. However, to increase the efficacy of medicinal plants identified, it is important to develop methods such as mixing with some fixative materials. Knowing that plant essential oils are promising alternatives to chemical insecticides [29] and have demonstrated their effectiveness against some Bostrichidae pests of stored products [30, 31], it is important to conduct research on their use on the control of *D. porcellus*.

3.3 Resistant varieties

The use of resistant varieties is the cheapest, effective, and ecologically safe method of protecting stored products against insect pests [32]. In fact, among the solutions proposed by farmers to fight *D. porcellus*, resistant yams chips have been the most plebiscite [10]. To meet the expectations of farmers, a participatory assessment of yam landraces in Benin was conducted in 51 villages through the yam production zone of Benin, and 64 landraces whose chips are resistant to storage insect have been identified [33]. The agro-morphological [33] and molecular characterization [34] of these 64 landraces revealed the existence of many duplicates and classified them in 24 morphotypes [33]. Choosing based on their good agronomic and culinary characteristics, the susceptibility of 5 of the 64 landraces identified as given resistant chips were tested in the laboratory by Onzo et al. [35]. These authors demonstrated the existence of a differential susceptibility between the different tested landraces with respect to *D. porcellus* with Singor and Portchahabim landraces as the least vulnerable to the attacks of this pest. Varietal resistance of yam chips from 24 landraces (corresponding to the 24 morphotypes obtained during morphological characterization) to *D. porcellus* was evaluated by Loko et al. [16] using free-choice tests (antixenosis) and non-choice tests under laboratory conditions. The results showed that basing on the Dobie index of susceptibility five yam landraces (Gaboubaba, Boniwouré, Alahina, Yakanougo, and Wonmangou) were scored as resistant to *D. porcellus*. These five resistant landraces (**Figure 3**) can be used in the integrated management of *D. porcellus*.

3.4 Biological control

Biological control is an important integrated pest management component (IPM), which broadly includes all control types involving living organisms, and represents a good alternative to the use of pesticides. Concerning *D. porcellus*, two predators (*Alloeocranum biannulipes Montrouzier* & Signoret (Hemiptera: Reduviidae) and *Teretrius nigrescens* Lewis (Coleoptera: Histeridae)) have been found in yam chips infested with this pest [10]. The functional responses of *T. nigrescens* and *A. biannulipes* feeding on *D. porcellus* were compared, and the results showed that both predators have a potential as biological control agents of *D. porcellus* [17]. The suppressive effect of *A. biannulipes* on the population dynamics of *D. porcellus* and the yam chip losses caused by this pest was evaluated under laboratory and natural conditions [18]. Results showed that *A. biannulipes* has the potential to be an effective biological agent against *D. porcellus* in stored yam chips (**Figure 4**). In addition, Loko et al. [18] provided detailed information on the biology, behavior, and life history of *A. biannulipes*, which are necessary for the mass rearing and use of this predator to control *D. porcellus*. However, *A. biannulipes* is a generalist predator which can consume several stored product insect pests such

Figure 3.
Pictures of chips from the five resistant landraces to D. porcellus identified by Loko et al. [16].

Figure 4.
Predation by A. biannulipes adult on nymph of D. porcellus.

as *Corcyra cephalonica* (Stainton) (Lepidoptera: Pyralidae), *Tribolium confusum* du Val (Coleoptera: Tenebrionidae), and *Anagasta kuehniella* Zeller (Lepidoptera: Pyralidae) [36]. What could be affected is its effectiveness as a biological agent of *D. porcellus* in farmer storage conditions. Therefore, prior to the use of this predator in an IPM program against *D. porcellus*, it is important to do a molecular gut analysis for determining the part of *D. porcellus* in its diet. Moreover, it is important to evaluate the population dynamics of *A. biannulipes* and *D. porcellus* within multispecies and/or multitrophic systems.

3.5 Synthetic insecticides

Synthetic insecticides are the main control method used by farmers and traders, to protect stored yam chips against insect attacks [7, 10]. In Benin, some recommended synthetic insecticides for storage insect pest control such as Sofagrain (1.5% deltamethrin +0.5% pirimiphos methyl) and Antouka (permethrin 3 g/kg + pirimiphos 16 g/kg) showed a good level of stored yam chip protection against *D. porcellus* [7, 10]. But in practice, most of the time, farmers use synthetic insecticides focusing other crops such as cotton in Benin [10] and cacao in Nigeria [37] to protect stored yam chips. This misuse of chemical insecticide leads to many cases of poisoning [13, 14]. Indeed, a study of Sosan et al. [37] revealed the presence of the organochlorine pesticide residues such as Dichloro-diphenyl-trichloro-ethane (DDT) and hexachlorohexane (HCH) at outrageous levels in dried yam chips obtained from Ile-Ife markets, southwestern Nigeria. These two organochlorine pesticide residues classified as dangerous by the World Health Organization [38] could cause serious health and environmental risks. Moreover, the use of aluminum phosphide marketed as Phostoxin, which is a highly toxic pesticide banned in several countries [39], was registered by Adesina et al. [40] as use by traders for yam chip protection. It is therefore important to sensitize farmers and traders to the use of chemical insecticides suitable for the protection of yam chips.

4. Integrated pest management

Integrated pest management (IPM) relies on managing insect populations through the combined use of several control methods in a way, which affords the highest priority to the protection of human health as well as the environment. The promising approaches toward effective IPM for *D. porcellus* are the use of resistant yam landraces combined with botanical powders of three medicinal plants (*B. ferruginea*, *B. sapida*, and *K. senegalensis*) [41]. Based on the few studies carried out in the context of the yam chip protection against *D. porcellus*, we can recommend to smallholder farmers the integration of the different methods in the following way:

- Use resistant landraces (Gaboubaba, Boniwouré, Alahina, Yakanougo, and Wonmangou) for yam chip manufacturing. These five yam landraces have a good agronomic (productivity, number of tubers), culinary (quality of pounded and boiled yam), and technological (quality of yam chips, ease of pounding) characteristics, found in Beninese traditional agriculture [16, 33].

- Peel and cut fresh yam tuber in slice ranging from 2 to 3 cm of thickness for fast drying [16].

- Precook at around 40°C, and soak during minimum 12 h fresh yam tubers before drying for having the best quality yam flour [42], but also protect chips against insect attacks [43]. Indeed, Nwana and Azodeh [44] showed that the intensity of damage by *A. fasciculatus* to yam chips blanched and soaked before drying were low.

- Add plants such as leaves and sorghum straw during the parboiling process for red coloration and its insect repellent properties [10]. Indeed, it is known that color largely affects the acceptability of Amala (thick paste obtained after mixed boil water and yam chips flour) by consumers [45].

- Dry yam chips on clean surface to avoid insect infestation that could inadvertently be carried to storage; yam chip moisture content must be less than 13% to avoid fungi and insect attacks [46].

- Put dried yam chips in polythene-lined jute for lower insect damage and yam chip discoloration [28].

- Add leaf powders of *B. ferruginea*, *B. sapida*, and *K. senegalensis* at a concentration of 5% (weight/weight) in bagged yam chips for a short period of conservation (3 months) because of their high repellent activity against *D.*

porcellus [15]. However, for a long period of yam chip conservation (10 months or more), we recommend the use of neem leaves (100 g per kg) or neem seeds (20 g per kg) powders [7].

- Tightly seal bags and put them in clean and dried place.
- Inspection of stored bags should be done monthly.
- If stored yam chips are attacked by *D. porcellus*, we recommend heating infested yam chips to more than 35°C [19] or adding the predator A. biannulipes at a density of one predator for 10 preys [18] or applying the recommended chemical insecticides Antouka or Sofagrain.

5. Scope for future research and development of innovative management strategies

In order to develop a good strategy of integrated pest management for *D. porcellus* in West Africa, further research should be done in the development of alternative control measures and techniques. The alternative methods to be explored that could be adapted to *D. porcellus* control in stored yam chips are as follows:

5.1 Hermetic bags

Hermetic storage bags have proven to be a low-cost solution for preventing storage losses due to insects [47]. Storage systems based on the hermetic principle can be used to maintain stored product quality without the need for pesticide application [48]. Apart from testing the effectiveness of the Purdue Improved Crop Storage (PICS bags™) in *D. porcellus* control, it is important to evaluate the efficacy of other hermetic methods of storage marketed in West Africa such as AgroZ Bag™, and SuperGrain bags™ [49].

5.2 Essential oils

Essential oils can have various effects on stored insect pests (repellence, contact toxicity, antifeedant, growth inhibitory, fumigant, etc.) and can be applied as a part of integrated pest management programs for stored products protection. Indeed, plant essential oils can be used in combination with other control techniques for controlling storage insect pest [50]. Essential oil of plants found in the West African flora such as *Citrus sinensis* [31], *Cymbopogon citratus* [51], *Ocimum basilicum* [52], and *Zingiber officinale* [53] have proven their efficacy on control of several Bostrichidae of stored products such as *P. truncatus* and *R. dominica*. Research on the potential use of essential oils of these plants or other medicinal plants found in West Africa in control of *D. porcellus* in stored yam chips should be prospected.

5.3 Entomopathogens

Entomopathogens have an important place in the biological control because they have a wide host range and are harmless to the environment and human. These include entomopathogenic fungi, nematodes, bacteria, and viruses. These are all widespread in the natural environment and cause infections in many pest species. Entomopathogens contribute to the natural regulation of many populations of arthropods. Much of the research in this area concerns the causal agents of insect diseases and their exploitation for biological pest control. Many entomopathogens can be mass produced, formulated, and applied to pest populations in an analogous manner to chemical pesticides. Also, they can be used more against stored product pests with the development of new biotechnical methods. Indeed, the effectiveness of the formulation of the entomopathogenic fungi, *Beauveria bassiana* (Bal.) Vuillemin and *Metarhizium anisopliae* (Metch.), against the Bostrichids *P. truncatus* [54, 55] and *R. dominica* [56, 57] has been proven. It would therefore be interesting to evaluate the effect of entomopathogenic fungi in the context of *D. porcellus* control.

5.4 Parasitoids

The parasitoid wasp, *Anisopteromalus calandrae* (Howard) (Hymenoptera: Pteromalidae) found in stored yam chips by Vernier et al. [7], is an important biological agent of several Bostrichidae larvae of stored products such as *R. dominica* [58], and *P. truncatus* [59]. Similarly, *Dinarmus basalis* Rondani (Hymenoptera: Pteromalidae), which is present in all the different regions of West Africa, was also found in stored yam chips [10]. However, *D. basalis* is known as an efficient ectoparasitoid of bruchid pests [60, 61]. Further research must be done to evaluate the potential of these two parasitoids as biological agents for the control of *D. porcellus* in stored yam chips.

5.5 Insect growth regulators

Insect growth regulators are insecticides that mimic hormones in young insects and can be a potential component in integrated pest management against *D. porcellus*. Several features of insect growth regulators make them attractive as alternatives to broad-spectrum insecticides. Indeed, insect growth regulators are more selective; they are less harmful to the environment and more compatible with pest management systems that include biological controls. In addition, insect growth regulators are generally low in toxicity to humans. Kavallieratos et al. [62] have proven the efficacy of insect growth regulators as grain protectants against *P. truncatus* in maize and *R. dominica* in wheat. Therefore, investigations must be done to assess the effects of insect growth regulators on the development of *D. porcellus*.

5.6 Pheromones

Methods for detecting and for monitoring *D. porcellus* are crucial components for the development of an integrated management strategy against this pest. Pheromones which are volatile organic molecules of low molecular weight that elicit a behavioral response from individuals of the same species can serve as a tool to detect infestation at an early stage and to determine the right time for control measures [63]. Thus, Campion et al. [64] and Hodges [65] have shown the efficiency of traps baited with the synthetic aggregation pheromones for detecting and monitoring *P. truncatus* in East and West Africa. Research should develop specific pheromones and attractants of *D. porcellus* to aid in its monitoring and trapping.

6. Conclusions

The pest status of *D. porcellus* is higher in West Africa. This pest causes both quantitative and qualitative damage to stored yam chips. Synthetic insecticides used by farmers and traders for fighting against this insect are very dangerous for human health and environment. The present chapter has emphasized utilization of resistant varieties, botanicals, parasitoids, and physical methods for *D. porcellus* control. Thus, more emphasis should be placed on the integration of these different methods of *D. porcellus* control. This chapter proposes an integrated pest management combining yam chip processing practices, physical methods, botanical insecticides, biological control, and resistant varieties for fighting *D. porcellus* in stored yam chips. However, detection and monitoring tools should be developed, and it appears that the use of hermetic bags, essential oils, entomopathogens, insect growth regulators, pheromones, and their combined effects should be further investigated in the *D. porcellus* control.

Conflict of interest

The authors declare that they have no competing interests.

Author details

Loko Yêyinou Laura Estelle*, Toffa Dèca Mondoukpè Joelle and Orobiyi Azize
Faculty of Sciences and Technology of Dassa (FAST-Dassa), Laboratory of Applied Entomology (LEnA), National University of Sciences, Technologies, Engineering and Mathematics of Abomey (UNSTIM), Dassa-Zoumé, Benin

*Address all correspondence to: lokoestelle@yahoo.fr

References

[1] FAO. FAOSTAT Database. Roma, Italy: Food and Agriculture Organization; 2017. Available from: www.fao.org [Accessed: 10 March 2019]

[2] Bakare HA, Adegunwa MO, Tossou HB, Durojaiye GD, Ibitayo FS, Tijani OA. Optimisation of the processing conditions on the culinary qualities of pressure-cooked boiled yam. Journal of Culinary Science and Technology. 2018:1-17. DOI: 10.1080/15428052.2018.1495589

[3] Olajumoke OL, Agiang MA, Mbeh E. Proximate and anti-nutrient composition of white Guinea yam (*Dioscorea rotundata*) diets consumed in Ibarapa, South West region of Nigeria. The Journal of Natural Product and Plant Resources. 2012;**2**(2):256-260

[4] Olorede KO, Alabi MA. Economic analysis and modelling of effects of NPK fertilizer levels on yield of yam. Mathematical Theory and Modeling. 2013;**3**(1):108-118

[5] Osunde ZD, Orhevba BA. Effects of storage conditions and storage period on nutritional and other qualities of stored yam (*Dioscorea* spp) tubers. African Journal of Food Agriculture Nutrition and Development. 2009;**9**(2): 678-690

[6] Umogbai VI. Design, construction and performance evaluation of an underground storage structure for yam tubers. International Journal of Scientific and Research Publications. 2013;**3**(5):1-7

[7] Vernier P, Goergen G, Dossou RA, Letourmy P, Chaume J. Utilization of biological insecticides for the protection of stored yam chips. Outlook on agriculture. 2005;**34**(3):173-179

[8] Hounhouigan JD, Akissoe N, Bricas N, Vernier P. Diagnostic des systèmes techniques de transformation de l'igname en cossettes séchées au Bénin. In: Berthaud J, Bricas N, Marchand JL, editors. L'igname, plante séculaire et culture d'avenir: Actes du séminaire international, Cirad, Inra, Orstom, Coraf, Cirad, Inra, Orstom, Coraf, Coll Colloques. CIRAD. 1998. pp. 349-351

[9] Babarinde SA, Babarinde GO, Odewole AF, Alagbe OO. Effect of the prevalent insect species of yam chips on consumers' acceptability of yam paste. Agricultura Tropica et Subtropica. 2013;**46**(3):97-101. DOI: 10.2478/ats-2013-0017

[10] Loko YL, Dansi A, Tamo M, Bokonon-Ganta AH, Assogba P, Dansi M, et al. Storage insects on yam chips and their traditional management in Northern Benin. The Scientific World Journal. 2013. Article ID: 484536. 11 p. DOI: 10.1155/2013/484536

[11] Adesuyi SA. The problems of insect infestation of stored yam chips in Nigeria. In: Proceedings of the Second International Working Conference on Stored-Product Entomology; 10-16 September 1978. Ibadan, Nigeria: IITA; 1979. pp. 314-319

[12] Osuji FNC. Observations on beetles attacking dried yams and yam flour from three Nigerian markets. Tropical Stored Products Information. 1980;**39**:35-38

[13] Adedoyin OT, Ojuawo A, Adesiyun OO, Mark F, Anigilage EA. Poisoning due to the yam flour consumption in five families in Ilorin, Central Nigeria. West African Journal of Medecine. 2008;**27**(1):41-43

[14] Adeleke SI. Food poisoning due to yam flour consumption in Kano (Northwest) Nigeria. The Online Journal of Health and Allied Sciences.

2009;**8**(2):10. Available from: http://www.ojhas.org/issue30/2009-2-10.htm

[15] Loko YL, Alagbe O, Dannon EA, Datinon B, Orobiyi A, Thomas-Odjo A, et al. Repellent effect and insecticidal activities of *Bridelia ferruginea*, *Blighia sapida*, and *Khaya senegalensis* leaves powders and extracts against *Dinoderus porcellus* in infested dried yam chips. 2017. Article ID: 5468202. 18 p. DOI: 10.1155/2017/5468202

[16] Loko YLE, Gnaho AC, Orobiyi A, Agre P, Dansi A, Tamò M. Resistance of dried chips of yam (*Dioscorea cayenensis-D. rotundata* complex) landraces to *Dinoderus porcellus* Lesne (Coleoptera: Bostrichidae). Cogent Food and Agriculture. 2017;**3**(1):1-12. DOI: 10.1080/23311932.2017.1411180

[17] Loko YL, Djagoun AD, Dannon EA, Datinon B, Dansi A, Thomas-Odjo AA, et al. Functional response of the predators *Alloeocranum biannulipes* (Hemiptera: Reduviidae) and *Teretrius nigrescens* (Coleoptera: Histeridae) feeding on *Dinoderus porcellus* (Coleoptera: Bostrichidae) infesting yam chips. Environmental Entomology. 2017;**46**(1):84-91

[18] Loko YLE, Gavoedo DM, Toffa J, Orobiyi A, Thomas-Odjo A, Tamò M. Life table of the predator *Alloeocranum biannulipes* Montrouzier and Signoret (Hemiptera: Reduviidae) and a test of its ability to suppress populations of *Dinoderus porcellus* Lesne (Coleoptera: Bostrichidae) in stored yam chips. Biological Control. 2019;**130**:60-69

[19] Oni M, Omoniyi A. Studies on temperature influence on oviposition and development of immature stages of the yam beetle *Dinoderus Porcellus* Lesne. Coleoptera: bostrichidae on dried yam species. The Journal of Agricultural Science. 2012;**4**(2):213-218

[20] Liu LY. Microstructural characters of Lyctinae and Dinoderinae (Coleoptera: Bostrichidae). Psyche. 2010. Article ID: 607568. 8 p. DOI: 10.1155/2010/607568

[21] Schäfer K, Goergen G, Borgemeister C. An illustrated identification key to four different species of adult Dinoderus (Coleoptera: Bostrichidae), commonly attacking dried cassava chips in West Africa. Journal of Stored Products Research. 2000;**36**:245-252

[22] Nwanna IE. A survey of storage Coleoptera which attack dried cocoyam chips in Nigeria. Journal of Stored Products Research. 1993;**29**:95-98. DOI: 10.1016/0022-474X(93)90028-3

[23] Banks, HJ; Fields, P. Physical methods for insect control in stored-grain ecosystems. In: Jayas D, White NDG, Muir WE, editors. Stored-Grain Ecosystems. New York, USA: Marcel Dekker; 1995. pp. 353-407.

[24] Chalal N, Bellhamri A, Bennamoun L. Etude d'un séchoir solaire fonctionnant en mode direct et indirect. Revue des Energies Renouvelables. SMSTS'08 Alger. 2008:117-126

[25] Sanon A, Dabiré-Binso LC, Ba NM. Triple-bagging of cowpeas within high density polyethylene bags to control the cowpea beetle *Callosobruchus maculatus* F. (Coleoptera: Bruchidae). Journal of Stored Products Research. 2011;**47**:210-215

[26] Hell K, Edoh Ognakossan K, Lamboni Y. PICS hermetic storage bags ineffective in controlling infestations of *Prostephanus truncatus* and Dinoderus spp. in traditional cassava chips. Journal of Stored Products Research. 2014;**58**:53-58. DOI: 10.1016/j.jspr.2014.03.003

[27] Hikal WM, Baeshen RS, Said-Al Ahl HAH. Botanical insecticide as

simple extractives for pest control. Cogent Biology. 2017;**3**(1):1-16. DOI: 10.1080/23312025.2017.1404274

[28] Eze SC, Asiegbu JE, Orkwor GC, Mbah BN, Asiedu R. Effect of four agrobotanical extracts and three types of bags on the control of insect pest and moulds of stored yam chips. Journal of Tropical Agriculture Food, Environment and Extension. 2006;**5**(1):8-12

[29] Campolo O, Giunti G, Russo A, Palmeri V, Zappalà L. Essential oils in stored product insect pest control. Journal of Food Quality. 2018. Article ID: 6906105. 18 p. DOI: 10.1155/2018/6906105

[30] Schmidt GH, Streloke M. Effect of *Acorus calamus* (L.) (Araceae) oil and its main compound β-asarone on *Prostephanus truncatus* (Horn) (Coleoptera: Bostrichidae). Journal of Stored Products Research. 1994;**30**(3):227-235. DOI: 10.1016/0022-474X(94)90050-R

[31] Campolo O, Romeo VF, Malacrinò A, Laudani F, Carpinteri G, Fabroni S, et al. Effects of inert dusts applied alone and in combination with sweet orange essential oil against *Rhyzopertha dominica* (Coleoptera: Bostrichidae) and wheat microbial population. Industrial Crops and Products. 2014;**61**:361-369. DOI: 10.1016/j.indcrop.2014.07.028

[32] Hiruy B, Getu E. Screening of some maize varieties for resistance against the Maize weevils, *Sitophilus zeamais* (Motsch.). International Journal of Entomology and Nematology. 2018;**4**(1):77-84

[33] Loko YL, Adjatin A, Dansi A, Vodouhè R, Sanni A. Participatory evaluation of Guinea yam (*Dioscorea cayenensis* Lam.– *D. rotundata* Poir. complex) landraces from Benin and agro-morphological characterization of cultivars tolerant to drought, high soil moisture and chips storage insects. Genetic Resources and Crop Evolution. 2015;**62**(8):1181-1192

[34] Loko YL, Bhattacharjee R, Paterne Agre A, Dossou-Aminon I, Orobiyi A, Djedatin GL, et al. Genetic diversity and relationship of Guinea yam (*Dioscorea cayenensis* Lam.–*D. rotundata* Poir. complex) germplasm in Benin (West Africa) using microsatellite markers. Genetic Resources and Crop Evolution. 2017;**64**:1205-1219

[35] Onzo A, Biaou JT, Loko LY, Tamo M, Dansi A. Vulnérabilité des cossettes issues de quelques cultivars d'igname à l'attaque de *Dinoderus porcellus* Lesne (Coleoptera: Bostrichidae) en conditions de laboratoire. International Journal of Biological and Chemical Sciences. 2015;**8**(6):2494-2507

[36] Awadallah KT, Tawfik FS, Abdellah MMH. Suppression effect of the reduviid predator, *Allaeocranum biannulipes* (Montr. et Sign.) on populations of some stored product insect pests. Zeitschrift für Angewandte Entomologie. 1984;**97**:249-253. DOI: 10.1111/j.1439-0418.1984.tb03745.x

[37] Sosan MB, Oyekunle JAO, Olufade YA. Dichloro-diphenyl-trichloro-ethane (DDT) and hexachlorohexane (HCH) pesticide residues in foodstuffs from markets in Ile-Ife, Nigeria. International Journal of Biological and Chemical Sciences. 2015;**9**(1):442-453

[38] World Health Organization (WHO). The WHO Recommended Classification of Pesticides by Hazard and Guidelines to Classification: 2004. Geneva: World Health Organization; 2005

[39] Moghadamnia AA. An update on toxicology of aluminum phosphide. DARU Journal of Pharmaceutical

Sciences. 2012;**20**:25. DOI: 10.1186/2008-2231-20-25

[40] Adesina GO, Babarinde SA, Olaniran AO. Assessment of selected food products for pesticide residue in major markets of Oyo state, Nigeria. International Letters of Chemistry, Physics and Astronomy. 2015;**54**:47-55

[41] Loko YLE, Gnaho CA, Toffa J, Orobiyi A, Dansi A, Tamo M. Management of *Dinoderus porcellus* L. (Coleoptera: Bostrichidae) infesting yam chips using varietal resistance and botanical powders of three medicinal plants. African Journal of Agricultural Research. 2018;**13**(40):2118-2133

[42] Adejumo BA, Okundare RO, Afolayan OI, Balogun SA. Quality attributes of yam flour (elubo) as affected by blanching water temperature and soaking time. The International Journal of Engineering and Science (IJES). 2013;**2**(1):216-221

[43] Gnonlonfin GJB, Hell K, Siame AB, Fandohan P. Infestation and population dynamics of insects on stored cassava and yams chips in Benin, West Africa. Journal of Economic Entomology. 2008;**101**(6):1967-1973. DOI: 10.1603/0022-0493-101.6.1967

[44] Nwana IE, Azodeh IC. The effect of variety and processing method on the damage to dried yams by *Araecerus fasciculatus* (Degeer) (Coleoptera: Anthribidae). Tropical Stored Product Information. 1984;**49**:3-7

[45] Chilaka FC, Eze S, Anyadiegwu C, Uvere P. Browning in processed yams: Peroxidase or polyphenol oxidase? Journal of the Science of Food and Agriculture. 2002;**82**:899-903

[46] Omohimi C, Piccirillo C, Ferraro V, Roriz MC, Omemu MA, Santos S, et al. Safety of yam-derived (*Dioscorea rotundata*) foodstuffs-chips, flakes and flour: Effect of processing and post-processing conditions. Food. 2019;**8**(1):12. DOI: 10.3390/foods8010012

[47] Lane B, Woloshuk C. Impact of storage environment on the efficacy of hermetic storage bags. Journal of Stored Products Research. 2017;**72**:83-89. DOI: 10.1016/j.jspr.2017.03.008

[48] Likhayo P, Bruce AY, Tefera T, Mueke J. Maize grain stored in hermetic bags: Effect of moisture and pest infestation on grain quality. Journal of Food Quality. 2018. Article ID: 2515698. 9 p. DOI: 10.1155/2018/2515698

[49] Obeng-Ofori D, Adarkwa C, Ulrichs C. Chemical, physical and organic hermetic storage technology for stored-product protection in African countries. In: Proceedings of the IOBC-WPRS Bulletin, Working Group "Integrated Protection of Stored Products"; June 28–July 1; Zagreb (Croatia); Vol. 111; 2015. pp. 3-27

[50] Ziaee M, Hamzavi F. A review of plant essential oils as a component of integrated pest management in stored products protection. In: Proceedings of International Conference on Green Agro-Industry (ICGAI); 12-14 November 2013. Indonesia: Yogyakarta; 2013. pp. 394-402

[51] Masamba WRL, Kamanula JFM, Henry EMT, Nyirenda GKC. Extraction and analysis of lemongrass (*Cymbopogon citratus*) oil: An essential oil with potential to control the larger grain borer (*Prostephanus truncatus*) in stored products in Malawi. Malawi Journal of Agricultural Sciences. 2003;**2**(1):56-64

[52] Kumar R. Repellent effect of essential oil from *Ocimum basilicum* against *Rhyzopertha dominica* (Coleoptera: Bostrichidae) and *Tribolium castaneum* (Coleoptera: Tenebrionidae). Journal of Postharvest Technology. 2016;**4**(3):S50-S52

[53] Ukeh DA. Bioactivities of essential oils of *Aframomum melegueta* and *Zingiber officinale* both (Zingiberaceae) against *Rhyzopertha dominica* (Fabricius). Journal of Entomology. 2008;**5**:193-199

[54] Meikle WG, Cherry AJ, Holst N, Hounna B, Markham RH. The effects of an entomopathogenic fungus, *Beauveria bassiana* (Balsamo) Vuillemin (Hyphomycetes), on *Prostephanus truncatus* (Horn) (Col.: Bostrichidae), *Sitophilus zeamais* Motschulsky (Col.: Curculionidae), and grain losses in stored maize in the Benin Republic. Journal of Invertebrate Pathology. 2001;**77**(3):198-205. DOI: 10.1006/jipa.2001.5015

[55] Kassa A, Zimmermann G, Stephan D, Vidal S. Susceptibility of *Sitophilus zeamais* (Motsch.) (Coleoptera: Curculionidae) and *Prostephanus truncatus* (Horn) (Coleoptera: Bostrichidae) to entomopathogenic fungi from Ethiopia. Biocontrol Science and Technology. 2002;**12**(6):727-736. DOI: 10.1080/0958315021000039905

[56] Batta YA. Control of the lesser grain borer (*Rhyzopertha dominica* (F.), Coleoptera: Bostrichidae) by treatments with residual formulations of *Metarhizium anisopliae* (Metschnikoff) Sorokin (Deuteromycotina: Hyphomycetes). Journal of Stored Products Research. 2005;**41**(2):221-229. DOI: 10.1016/j.jspr.2004.03.007

[57] Mahdneshin Z, Safaralizadah MH, Ghosta Y. Study on the efficacy of Iranian isolates of *Beauveria bassiana* (Balsamo) Vuillemin and *Metarhizium anisopliae* (Metsch.) Sorokin against *Rhyzopertha dominica* F. (Coleoptera: Bostrichidae). Journal of Biological Sciences. 2009;**9**:170-174. DOI: 10.3923/jbs.2009.170.174

[58] Menon A, Flinn PW, Dover BA. Influence of temperature on the functional response of *Anisopteromalus calandrae* (Hymenoptera: Pteromalidae), a parasitoid of *Rhyzopertha dominica* (Coleoptera: Bostrichidae). Journal of Stored Products Research. 2002;**38**(5):463-469. DOI: 10.1016/S0022-474X(01)00050-9

[59] Helbig J. Ability of naturally occurring parasitoids to suppress the introduced pest *Prostephanus truncatus* (Horn) (Coleoptera, Bostrichidae) in traditional maize stores in Togo. Journal of Stored Products Research. 1998;**34**(4):287-295. DOI: 10.1016/S0022-474X(98)00010-1

[60] Sanon A, Ouedraogo AP, Tricault Y, Credland PF, Huignard J. Biological control of Bruchids in cowpea stores by release of *Dinarmus basalis* (Hymenoptera: Pteromalidae) adults. Environmental Entomology. 1998;**27**(3):717-725. DO: 10.1093/ee/27.3.717

[61] Hossain MA, Alim MA, Ahmed KS, Haque MA. Biocontrol potential of *Dinarmus basalis* (Pteromalidae: Hymenoptera) Rondani as a parasitoid of *Callosobruchus chinensis* (L.) in stored pulse. African Entomology: Journal of the Entomological Society of Southern Africa. 2014;**22**(2):285-290

[62] Kavallicratos NG, Athanassiou CG, Vayias BJ, Tomanović Z. Efficacy of insect growth regulators as grain protectants against two stored-product pests in wheat and maize. Journal of Food Protection. 2012;**75**(5):942-950. DOI: 10.4315/0362-028X.JFP-11-397

[63] Landolt PJ, Phillips TW. Host plant influences on sex pheromone behavior of phytophagous insects. Annual Review of Entomology. 1997;**42**:371-391. DOI: 10.1146/annurev.ento.42.1.371

[64] Campion DG, Hall DR, Prevett PF. Use of pheromones in crop and stored

products pest management: control and monitoring. Recent Advances in Research on Tropical Entomology. 2011;**8**:737-741. DOI: 10.1017/S1742758400022852

[65] Hodges RJ. Detection and monitoring of larger grain borer, *Prostephanus truncatus* (Horn) (Coleoptera: Bostrichidae). Integrated Pest Management Reviews. 2002;7(4):223-243. DOI: 10.1023/b:ipmr.0000040815.06804

Chapter 6

Influence of Temperature and Storage Systems on Post-Harvest Losses of Maize Varieties Cultivated at Alibori in Northern Benin

Corinne M. Anagonou, Roland Dossou, Anicet G. Dassou and Alexandre Dansi

Abstract

Majority of post-harvest losses of several maize varieties observed in various storage systems in northern Benin are mainly caused by storage insects due to changes in climatic parameters. The objective of this study is to evaluate the levels of insect pest infestation of three maize varieties stored in storage systems at different temperature. In 18 villages at Alibori, maize farmers were surveyed through a participatory research approach and their storage structures were also visited. The temperature of all storage structures were noted. Weight loss of samples, numbers of *Prostephanus truncates*, *Sitophilus zeamais* and perforated grains were evaluated. In total, three maize varieties and three different groups of storage systems were identified during field observations. All the three maize varieties stored in the first storage systems group built with plants were less infested and had acceptable nutritional quality than the maize grains stored in the second group built in banco and third group built with tarpaulin. In these storage systems, the yellow maize variety was the most attacked, followed by the white maize variety and finally the mixed color of yellow and white maize variety the less attacked. Effective post-harvest management of stored products requires clear monitoring criteria of climatic parameters and effective implementation of abiotic and biotic factors.

Keywords: maize, storage system, insect pests, post-harvest losses, temperature, Alibori

1. Introduction

Maize is the basic food in most of developing countries [1]. Maize is also important for commercial transactions [2]. To increase agricultural income, have good quality of seeds and ensure permanent availability of maize in the market over a long period, farmers use different post-harvest storage systems to conserve the maize. In Eastern Senegal, the storage of maize grains is done in bags, racks, granaries, barrels, shops and others [3]. In Tanzania, in polyethylene bags, granaries, cans and other plastic containers are also used for storing maize [4]. In Benin, maize is stored in traditional granaries built from straw, bamboo, branches or reeds used to store spathed or despathed ears; in earth granaries for maize grain storage;

in artisanal cribs; in stores for large maize farmers (often in 100 kg bags). [5] had to distinguish two forms of granary: the traditional granaries (the type "Ago" and the type "Ava") and improved granaries made in plant materials and earth closed.

Majority of these storage systems have enough post-harvest losses often recorded in stored maize. More than 30% of grain crop harvests including maize are lost during storage in sub-Saharan Africa [6]. Maize post-harvest losses in tropics in general [7] and especially in Benin [8] can reach 40% after five months of storage. These losses are mainly due to the pests attack associated with the variation of temperature in the granaries. The most common storage insects are *Prostephanus truncatus* (Horn) (Coleoptera: Bostrichidae), *Rhyzopertha dominica* (F.) (Coleoptera: Bostrichidae), and the *Sitophilus zeamais* (Coleoptera: Curculionidae) [9–11]. From these storage insects, *Prostephanus truncatus* (Horn) and *Sitophilus zeamais* (Motschulsky) are the major insects observed in maize stocks [8, 12]. In rural areas where conservation techniques are poorly developed, *S. zeamais* can cause post-harvest losses of up to 90% for five months of storage [13, 14]. They cause damage including weight loss, a decrease in grain quality [15] and sometimes a loss of germination [16].

Defective storage structures with uncontrolled climatic conditions increase the abundance of storage insects and their damage in stored products. Many traditional granaries are not well ventilated and maintained in optimal temperature for the development of storage insects. Keeping maize grains in storage structures with low temperature and humidity helps to reduce the damage caused by storage insects [17]. Few studies emphasized on the temperature and humidity at which the grains of maize must be stored in storage structures to reduce the damage of storage insects.

Therefore, this study was conducted in Alibori region to identify the endogenous storage systems used by farmers for maize conservation and to evaluate the losses caused by these storage structures in various climatic conditions. The main objective of this research is to evaluate the influence of maize storage systems with different temperature on maize post-harvest losses in Alibori Region. Specifically, the present study aims to: (i) Assess the effect of yellow, white and mixed color of yellow and white maize varieties on post-harvest losses to determine which maize varieties cultivated in the study area are more resistant to insects attack; and (ii) Determine the influence of internal temperature inside storage structures on maize post-harvest losses to suggest to farmers the best post-harvest structures adapted to the better maize conservation.

2. Methodology

2.1 Study area

Alibori is one of Northern Benin region. It is located between 11°19′ north latitude and 2°55′ east longitude. It is bordered in the North by the Republic of Niger, in the North-West by the Republic of Burkina Faso, in the East by the Federal Republic of Nigeria, in the West by the Atacora and in the South by the Borgou Region (**Figure 1**). The daily temperature varies between 22° to 40°C. With an area of 26,242 km^2 (23% of the national territory), Alibori is subdivided into six Municipalities which are Malanville, Karimama, Ségbana, Gogounou, Banikoara and Kandi, making up 41 districts and 229 villages. Its population is estimated at 867,463 inhabitants. The climate is of the Sudanian type in its southern part and Sudano-sahelian in its northern part (Karimama and Malanville). There is only one season of rain which lasts between 5 and 6 months with an oscillating rainfall

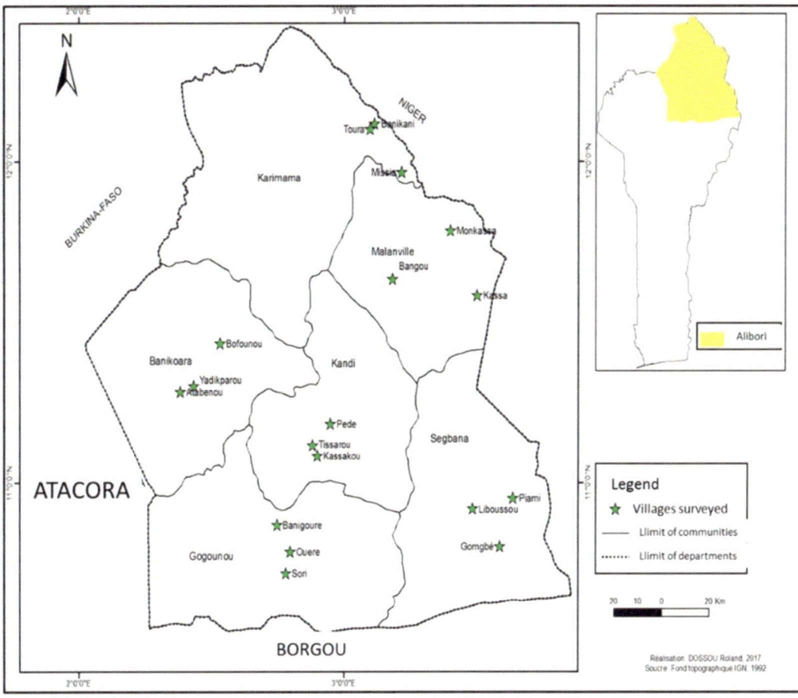

Figure 1.
Presentation of the study area and geographical position of the surveyed villages.

between 700 and 1200 mm. The vegetation is composed of a sparse shrub savannah, dominated by thorny trees, including *Acacia seyal* and *Acacia siebenona* in the north; and a grassy savannah heavily degraded to the South.

2.2 Data sampling

Data were collected in 18 villages during the months of April to June at the different maize storage sites through the application of participatory research tools and techniques such as direct observation, individual interviews and field visits using a questionnaire [18]. To identify the storage structures in the study area, farmers were asked to give the name of the storage structures or storage used by their household. Subsequently, these storage structures were visited and photographed for better description. Probe Thermometers were introduced at different places in these maize storage structures to note daily the internal ambient temperature. To assess the maize post-harvest losses caused by storage insects in the study during storage, approximately 1 kg of maize grains and maize corns of three varieties (white, yellow and white-yellow) was collected from all storage structures. A total of six samples including three varieties in the form of grains and also three in the form of corns were collected in each structure. The three storage structures such as granaries in banco, granaries in Plant Materials and conservation with Tarpaulin were used in the study. Daily temperature values were recorded in each storage structure during survey periods using metal probe thermometers. In each storage structure, probe thermometer was placed in three different locations such as at the roof, at the base and on side. The daily temperature was obtained by calculating the average value of the three temperature measurements made on each type of granary. All farmers store their products during the same period after the rainy season. Concerning the evaluation of storage losses, the initial and final weight of maize samples, the number and weight of the perforated maize grains were evaluated in the laboratory. Insect densities were calculated per kilogram of maize grains.

2.3 Data analyses

We used a Generalized Linear Model (GLM) with the family binomial and Analysis of Variance (ANOVA) to determine the effects of storage structures, and forms of conservation of maize varieties in (i) proportions of pest damage and (ii) densities of *Prostephanus truncates* and *Sitophilus zeamais*. The same analyzes were performed using the Generalized Linear Model (GLM) with the family Poisson and Analysis of Variance (ANOVA) to determine the effects of temperature of different storage structures on the masses of grains damaged. The test of Tukey HSD was used to determine the difference of masses of grains damaged between the storage structures. All the analysis was performed with the statistical software R version 3.4.2 [19].

3. Results

3.1 Densities of storage insects and effects of storage structures, and maize varieties on the insect pest damage

Prostephanus truncatus and *Sitophilus zeamais* were the two most abundant pests in maize stocks and their densities varied by communities ($P < 0.00001$). The density of *Prostepahanus trucatus* alive was higher in Malanville followed by Kandi and then the other municipalities. In contrast, the density of *Sitophilus zemais* was higher in Segbana followed by other municipalities (**Table 1**). Concerning the losses in number of damaged maize grains, the damage is observed much in Kandi followed by Malanville.

Even, all farmers noted that *Prostephanus truncatus* and *Sitophilus zeamais* were both major storage insects damaging the maize in post-harvest systems in study areas and it was confirmed by our observations. Post-harvest losses by volume were the most identified by the majority of farmers in all the study communities compared to post-harvest losses by weight. Municipalities of Gogounou and Kandi had the highest percentages of farmer responses in terms of volume and weight. Losses (**Figure 2**).

According to maize varieties, a significant effect was noted in maize weight loss and was positive for the yellow variety showing that post-harvest losses were enormous in this variety. A negative effect was observed for the mixed color of yellow and white maize variety showing that losses were reduced in this variety (**Table 2**). The boxplot carrying out the relationship between the maize varieties and weight of damaged maize grains by storage insects has shown that the yellow variety was the most attacked, followed by the white variety and finally the mixed color of yellow and white variety was the least attacked (**Figure 3**). The analysis of variance (ANOVA) showed a significant effect of the structures and forms of storage on the number of damaged maize grains ($Df = 2$, $P < 0.00001$). The test of Tukey HSD showed a significant difference only between the three maize varieties ($P < 0.00001$) and the three storage structures ($P < 0.00001$) for the infestation percentages.

3.2 Effects of temperature within different storage structures on the weight of grains damaged

The temperature of different storage structures had a significant and positive effect on damaged maize kernels weight ($P < 0.00001$, z-value = 6.21). Granaries built in plant materials conserved a low temperature and presented the low

Pests/damages	Banikoara	Gogounou	Kandi	Karimama	Malanville	Segbana
Live Prostephanus truncatus	3.3 ± 0.08	4.12 ± 0.13	6.5 ± 1.99	3.5 ± 0.04	9.5 ± 0.68	5.57 ± 2.13
Dead Prostephanus truncatus	0.4 ± 0.04	0.25 ± 0.06	0.6 ± 0.01	0.00 ± 0.00	0.00 ± 0.00	0.28 ± 0.08
Live Sitophilus zeamais	2.1 ± 0.027	3.5 ± 0.32	1.5 ± 0.76	0.5 ± 0.01	0.5 ± 0.06	11.57 ± 0.018
Dead Sitophilus zeamais	0.8 ± 0.122	0.125 ± 0.04	0.1 ± 0.01	0.00 ± 0.00	0.00 ± 0.00	2.14 ± 0.47
Number of damaged maize grains	3.4 ± 0.12	4.125 ± 0.27	6.8 ± 0.18	0.50 ± 0.00	4.62 ± 0.13	4.14 ± 2.54
Weight of damaged maize grains	0.77 ± 0.34	0.63 ± 0.07	1.24 ± 0.09	0.17 ± 0.03	1.21 ± 0.17	0.96 ± 0.058

Table 1.
Densities of storage insects and damaged maize grains according to study communities.

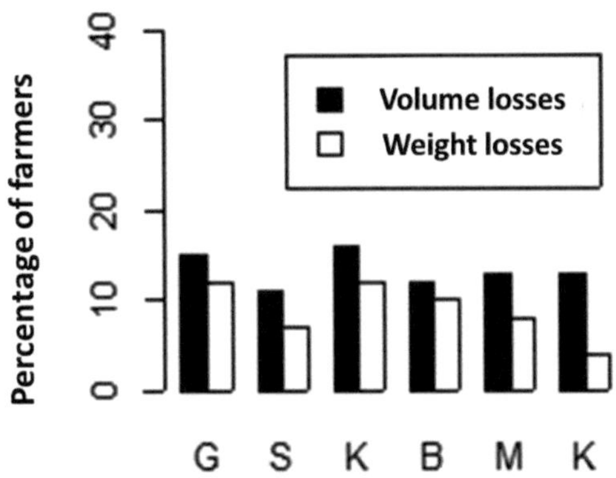

Figure 2.
Frequency of farmers listing the different types of post-harvest losses by communities. G, Gogounou; S, Ségbana; K, Kandi; Banikoara M, Malanville and K, Karimama.

Damage	Maize varieties	Df	Estimates	z-value	Pr (>|z|)
Number of damaged maize grains	Yellow	2	−1.5268	−4.270	<0.00001
	Mixed color (yellow- white)	2	−0.0826	−0.090	0.928 ns
	White	2	−1.0306	−0.563	0.574 ns
Weight of damaged maize grains	Yellow	2	1.46634	18.315	<0.00001
	Mixed color (yellow- white)	2	−0.77319	−2.771	0.005582
	White	2	0.64388	3.360	0.000778

Table 2.
Effect of maize varieties on the number and weight of damaged maize grains.

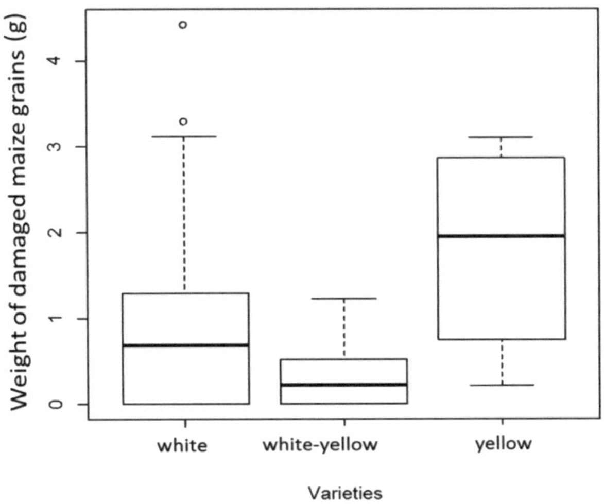

Figure 3.
Boxplot showing the level of weight loss according to the maize varieties.

damaged maize grains weight while the storage structures made with the tarpaulin protection presented high temperature and infestation (**Figures 4** and 5). The test of Tukey HSD showed a significant difference only between the three storage structures (P < 0.00001) for the temperature.

Figure 4.
Some storage structures of maize built in study areas.

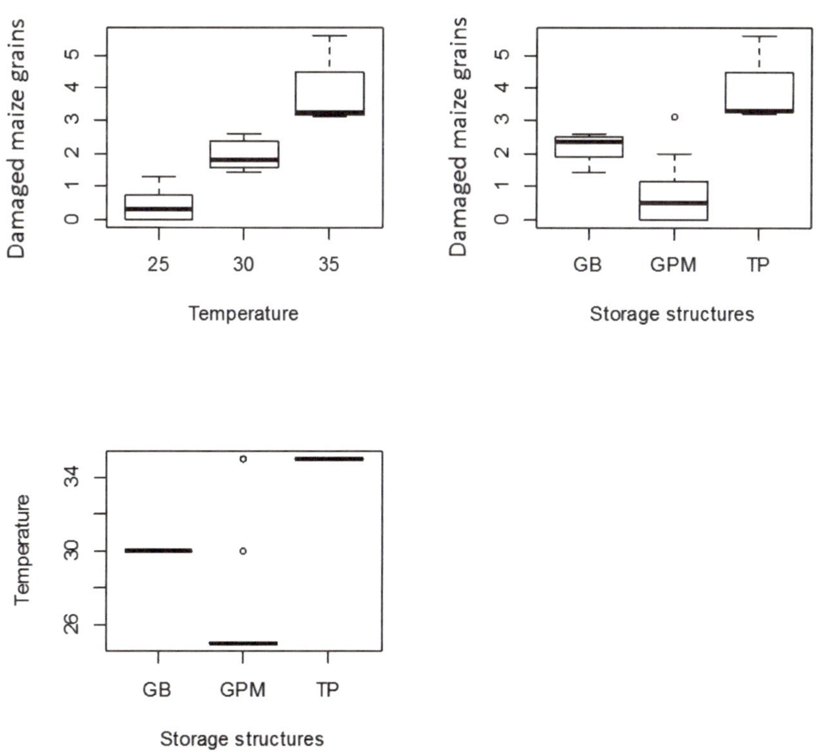

Figure 5.
Relationship between temperature of different storage structures and maize post-harvest losses. GB, granaries in banco, GPM, granaries in plant materials; TP, conservation with tarpaulin protection.

4. Discussion

4.1 Densities of storage insects and influence of storage structures and maize varieties on the insect pest damage

The main pests detected in laboratories on maize samples collected in the study area were *Prostephanus truncatus* followed by *Sitophilus zeamais*. Similar study conducted by [3, 4, 6, 20] had already recognized these insects as insects pests with huge losses of maize stocks. The abundance of certain storage insects would have a significant effect according to municipalities. This variation in insect densities can be justified by the singularity and the variation of the storage systems from one community to another. These recorded weight losses can also be justified by the use of traditional loft in banco (in Segbana), uncemented chamber (in Malanville) and bags. The post-harvest losses evaluated in terms of the number of damaged grains are more observed in maize samples from Gogounou Malanville and Segbana communities than in other communities. These represent the major production areas of maize in Alibori region and therefore include many traditional storage systems.

Additionally, the results obtained show that during maize storage, farmers have more post-harvest losses in volume than post-harvest losses in weight. This can be explained by the fact that storage structures used in study areas favor the loss of volume more than the weight losses. Statistical analyzes revealed that the form of maize stored has no significant effect in the post-harvest losses. This shows that all the storage systems encountered in the study environment have a storage defect due to their construction.; Only the level of insects attack varies from one structure to another [21]. The result are consistent with those of [3] in Senegal which reported that post-harvest losses of corn are independent of the mode or form of maize stored.

Furthermore, statistical analyzes revealed that post-harvest losses evaluated by weight do not depend on the communities. This can be explained by the fact that all the maize samples collected in all the villages of the six Municipalities are sensitive to post-harvest losses evaluated by weight [22]. These recorded weight losses can also be justified by the use of traditional granaries in banco, uncemented storage and bags. The results also showed that the yellow maize variety favors the development of storage insects as the white maize variety. It means that insects attack differs according to the maize varieties. First of all, the yellow variety wasthe most damaged followed by the white maize variety and the mixed color of yellow and white maize variety respectively.

4.2 Effects of temperature within different storage structures on the weight of maize grains damaged

The results showed that in the average temperature of 30–35°C in storage structures, the damage of storage insects on stored maize is high. This shows that these temperatures are optimal for rapid reproduction of storage insects in storage systems in the study area. Storage structures with internal temperatures of 30–35°C may be improved to reduce the damage of storage insects to the stored maize [17]. Other studies have shown that an average temperature of 30°C is optimal for the proliferation of storage insects and that above 35°C the temperature becomes lethal to these insects [12, 17, 23]. The use of tarpaulins and banco storage structures should be discouraged to producers for better conservation of maize. On the other hand, the results showed that at an average temperature of 25°C, maize grains were well preserved in storage structures made with plant materials and has low infestations. Further studies have been conducted to show that low temperature (<15° C)

DOI: http://dx.doi.org/10.5772/intechopen.88188

increases mortality and reduces oviposition and fecundity of *S. zeamais* and *S. oryzae* [24, 25]. Storage structures made from plant materials have good aeration and maintain a low internal temperature reducing the development of storage insects.

5. Conclusion

The different maize storage systems encountered in the study area have influence on post-harvest losses. Although grain storage is the preferred mode in the study area, it is more attacked by storage insects than corn on the cob. In addition, the 100 kg bags and banco granaries used by corn farmers in the study area were the storage structures that favored the attack of storage insects. The yellow variety was the most attacked followed by the white variety and the yellow-white respectively.

Acknowledgements

We thank maize farmers in Northern Benin who unconditionally accepted to respond to interviews and make their fields and maize storage structures available for observations.

Conflict of interest

The authors declare that they have no conflict of interest.

Author details

Corinne M. Anagonou[1,2], Roland Dossou[1], Anicet G. Dassou[1]* and Alexandre Dansi[1]

1 Laboratory of Biotechnology, Genetic Resources and Plant and Animal Breeding, National University of Sciences, Technologies, Engineering and Mathematics, Benin

2 Centre of Excellence for Agriculture Development and Sustainable Environment, Federal University of Agriculture of Abeokuta, Nigeria

*Address all correspondence to: dassoua5@gmail.com

References

[1] Hounsou CM, Arodokoun DY, Sikirou R, Zannou ET, Bello S, Afoha SAP, et al. Efficacité de l'Actellic Gold Dust Dp, insecticide binaire à base de Pyrimiphos-Méthyl et de Thiamethoxam, pour la lutte contre *Sitophilus zeamais* dans les stocks de maïs au Bénin; 2015

[2] Maboudou AG, Adégbola PY, Coulibaly O, Hell K, Amouzou ME. Factors affecting the use of improved clay store for maize storage in the central and northern Benin: 12-18. In: Fischer T, editor. New Directions for a Diverse Planet. Proceedings of the 4th International Crop Science Congress. Brisbane, Australia; 2004, 2004

[3] Guèye MT, Seck D, Wathelet JP, Lognay G. Lutte contre les ravageurs des stocks de céréales et de légumineuses au Sénégal et en Afrique occidentale: synthèse bibliographique. Biotechnologie, Agronomie, Société et Environnement (Biotechnology, Agronomy, Society and Environment). 2011;**15**(1):183-194

[4] Adebayo RA, Ibikunle O. Potentials of rice husk ash, cowdung ash and powdered clay as grain protectants against *Callosobruchus maculatus* (F) and *Sitophilus zeamais* (Mots). Applied Tropical Agriculture. 2014;**19**(2):48-53

[5] Arouna A, Adegbola PY. Analyse de la rentabilité financière des systèmes de stockage et de conservation du maïs au Sud-Bénin. Bulletin de la Recherche Agronomique du Benin: Numero Spécial. 2011;**2**:24-32

[6] Ngamo LST, Hance T. Diversité des ravageurs des denrées et méthodes alternatives de lutte en milieu tropical. Tropicultura. 2007;**25**(4):215-220

[7] Laberyrie V. Problème fondamentaux posés par les insectes des denrées. In: Conférence inaugurale. La post-récolte en Afrique. Actes du séminaire international tenu à Abidjan. Côte-d'Ivoire; 1991. pp. 9-14

[8] Kossou DK. Comparison of improved and local maize varieties in the Republic of Benin with emphasis on susceptibility to *Sitophilus zeamais* (Motschulsky). Journal of Stored Products Research. 1993;**29**:333-343

[9] Hodges RJ, Halid H, Rees DP, Meik J, Sarjono J. Insect traps tested as an aid to pest management in milled rice stores. Journal of Stored Products Research. 1985;**21**(4):215-229

[10] Kossou DK, Aho N. Stockage et conservation des grains alimentaires tropicaux. Principes et Pratiques. Cotonou, Bénin: Les éditions du Flamboyant; 1993. pp. 47-125

[11] Faye A, Thiaw C, Gueye-Ndiaye A, Sembène M. First investigation of different Crateava religiosa Forst formulations on the cowpea (Vigna unguiculata Walp.) seed-beetle, *Callosobruchus maculatus* Fabricius. International journal of science and advanced. Technology. 2012;**2**(8):56-65

[12] Harein PK, Press AF Jr. Mortality of stored-peanut insects exposed to mixtures of atmospheric gases at various temperatures. Journal of Stored Products Research. 1968;**4**(1):77-82

[13] Nukenine EN, Adler C, Reichmuth C. Efficacy of *Clausena anisata* and *Plectranthus glandulosus* leaf powder against *Prostephanus truncatus* (Coleoptera: Bostrichidae) and two strains of *Sitophilus zeamais* (Coleoptera: Curculionidae) on maize. Journal of Pest Science. 2010;**83**(2):181-190

[14] Baoua IB, Amadou L, Ousmane B, Baributsa D, Murdock LL. PICS bags for post-harvest storage of maize grain in

West Africa. Journal of Stored Products Research. 2014;**58**:20-28

[15] Rajendran S, Sriranjini V. Plant products as fumigants for stored-product insect control. Journal of Stored Products Research. 2008;**44**(2):126-135

[16] Dabire CL, Niango Ba M, Sanon A. Effects of crushed fresh Cleome viscosa L.(Capparaceae) plants on the cowpea storage pest, *Callosobruchus maculatus* fab. (Coleoptera: Bruchidae). International Journal of Pest Management. 2008;**54**(4):319-326

[17] Fields P, Korunic Z. The effect of grain moisture content and temperature on the efficacy of diatomaceous earths from different geographical locations against stored-product beetles. Journal of Stored Products Research. 2000;**36**(1):1-13

[18] Gbèdolo AE, Dassou AG, Dassou HG, Aminon ID, Omondi BA, Dansi A. Morphotype diversity of *Corchorus olitorius* and influence of agricultural practices on its potential major pest insects. Scientia Horticulturae. 2018;**239**:234-241

[19] R Development Core Team. R: A Language and Environment for Statistical Computing Computer Program, version By R Development Core Team, Vienna, Austria; 2014

[20] Dendy J, Dobie P, Saidi JA, Smith JL, Uronu B. Trapping the larger grain borer *Prostephanus truncatus* in maize fields using synthetic pheromones. Entomologia Experimentalis et Applicata. 1989;**50**(3):241-244

[21] Abass AB, Ndunguru G, Mamiro P, Alenkhe B, Mlingi N, Bekunda M. Post-harvest food losses in a maize-based farming system of semi-arid savannah area of Tanzania. Journal of Stored Products Research. 2014;**57**:49-57

[22] Affognon H, Mutungi C, Sanginga P, Borgemeister C. Unpacking postharvest losses in sub-Saharan Africa: A meta-analysis. World Development. 2015;**66**:49-68

[23] Howe RW. A summary of estimates of optimal and minimal conditions for population increase of some stored products insects. Journal of Stored Products Research. 1965;**1**(2):177-184

[24] Nakakita H, Ikenaga H. Action of low temperature on physiology of *Sitophilus zeamais* Motschulsky and *Sitophilus oryzae* (L.)(Coleoptera: Curculionidae) in rice storage. Journal of Stored Products Research. 1997;**33**(1):31-38

[25] Ileleji KE, Maier DE, Woloshuk CP. Evaluation of different temperature management strategies for suppression of *Sitophilus zeamais* (Motschulsky) in stored maize. Journal of Stored Products Research. 2007;**43**(4):480-488